THE AIR FORCE MUSEUM

This book is a pictorial tribute to the United States Air Force Museum which traces its history back to 1923 and to the history of controlled, powered, and sustained flight which traces back to the 1903 accomplishments of the Wright brothers of Dayton, Ohio.

In the original foreword to this book in 1975, then Lt. Gen. Ira C. Eaker noted that "Americans have been fascinated with airplanes, new and old, since World War I. That interest accounts for our preeminence in aerospace from the dawn of flight to the first human footprints on the moon. As long as that interest continues, we shall hold our air and space leadership."

The U.S. Air Force Museum helps to foster that interest by presenting historical and educational exhibits that tell the story of the United States Air Force from the days of the Wright brothers to its role in world history today. It should be noted here that the late General Eaker had a major role in aviation history as a pioneer aviator, organizer and commander of the Eighth Air Force in WW II, and in later civilian life as an advocate for aerospace development. For his lifetime achievements, in 1979 he was awarded a Special Medal of Honor. In 1985, in a special Pentagon ceremony, he was promoted to four-star general.

By presenting the story of the Air Force to its many visitors, the U.S. Air Force Museum also helps to develop a sense of national pride. This goal was underscored in the 1960s by the late Eugene W. Kettering, first chairman of the board of the Air Force Museum Foundation Inc., which has financed most of the construction of the current museum. Kettering wrote: "One of man's most noble emotions is patriotism—a deeply sincere and reverent love for his country. Unfortunately, patriotism is not innate, but must be taught to and learned by the individual."

Through this unique book, we attempt to carry forward the goals envisioned for the U.S. Air Force Museum by these distinguished Americans. And as General Eaker wrote, "This book...is next best to visiting the Air Force Museum." Thus, we encourage you to do just that in detail while the museum continues to commemorate what the Wright brothers accomplished for the nation and the world in 1903.

The Authors

D1222785

Copyright © 2003, 1999, 1991, 1986, 1983, 1980, 1977, 1975 by Nick P. Apple and Gene Gurney
All rights reserved. No part of this book may be reproduced or utilized in any form or by any means, electronic or
mechanical, including photocopying, digital, recording, or by any information storage and retrieval system without
permission in writing from the Publisher. Printed for publishers, Nick P. Apple and Gene Gurney, by the Central
Printing Company, Dayton, Ohio.
Manufactured in the United States of America

Library of Congress Cataloging-in-Publication Data

Apple, Nick P
  The Air Force Museum.

  Includes index.
  1. United States. Air Force Central Museum,
Wright-Patterson Air Force Base, Ohio. 2. Aeronautics,
Military — History. I. Gurney, Gene, joint author.
II. Title.
UG623.3.U62D3832   1975   358.4'0074'017173   75-8670   77-17895
ISBN 0-517-52018-4   ISBN 0-517-56296-0 (cloth)   0-517-56297-0 (paper)

THE AIR FORCE MUSEUM

Eighth Revised Edition

By
Lt. Col. Nick P. Apple
and Lt. Col. Gene Gurney

CENTRAL PRINTING
A Hammer Graphics Company

Front cover design by
Jerry Rep
AF Museum Foundation

The United States Air Force Museum

Before 2002

The main complex of the U.S. Air Force Museum dominates this portion of historic Wright Field, also known as Area B of Wright-Patterson Air Force Base. In addition, museum aircraft are displayed in the twin-hangars (*top center*). Here are the Presidential Aircraft Hangar (*with Wright brothers likeness*) and the Research and Development/Flight Test Hangar. Directly behind them, in Building 5, are located the Collections, Exhibits, and Research Divisions. The Restoration Division is in Hangars 4C-E out of view to the right. Aircraft also are displayed in the Outdoor Air Park to the left of the main complex. The museum's Memorial Park is shown to the right. Ample free parking is available in designated areas.

The Air Force Museum
Today and Tomorrow

"Reaching New Heights"

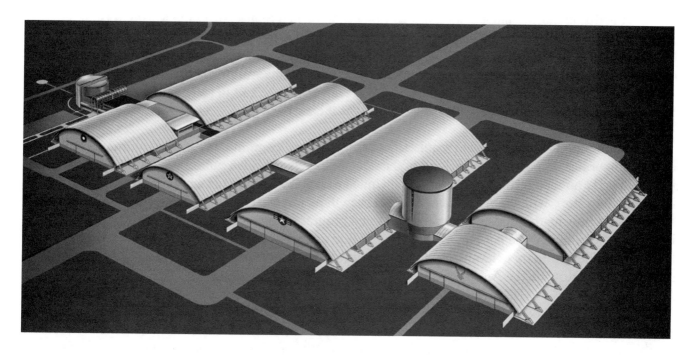

Prior to 2002, the southwest bay of the first building housed the Early Years Gallery featuring the history of the Air Force from the days of the Wright brothers to the pre-Pearl Harbor era. The larger northeast bay housed the Air Power Gallery featuring World War II, Korean and Vietnam Conflicts, and the Space Gallery. The second building, the Modern Flight Hangar, displayed a variety of aircraft from the past six decades, including the Cold War, rescue, and experimental aircraft. With the growth of the museum, the northeast bay is to concentrate on World War II while the second building is to feature aircraft and related exhibits from the Korean War, Southeast Asia War, and the more modern era. The Early Years Gallery carries on with its traditional role.

The third hangar-like structure is slated for completion by mid-2003. It will be the Eugene W. Kettering Building housing the Cold War Gallery and covering more than four and one-half acres of display space. Next to the Kettering Building, a Hall of Missiles will house the vertical Cold War missiles that formerly were displayed of necessity outdoors and thus unprotected from the weather. Temporarily, some Space Gallery items are to be displayed in the new Kettering Building.

Finally, a fourth hangar-like structure is to be built in two sections. The larger section would present the Air Force's dynamic role in the nation's journey into space. The other section would add classrooms for further enrichment of the museum's extensive educational program for both children and adults. As illustrated above, all the structures would be interconnected. And while the various construction projects are underway, visitors to the museum--and astute readers of this edition--may well discover that some attractions have been renamed and that various aircraft and exhibits may be in temporary locations as the dynamic museum continues its growth.

Individuals and organizations wishing to provide financial support for this expansion should contact the Air Force Museum Foundation Inc. at PO Box 33624, Wright-Patterson AFB, OH 45433-0624, or telephone (937) 258-1218, or fax (937) 258-3816.

★ Introduction

It is fitting that we launch our latest edition of the United States Air Force Museum Book in conjunction with the 100th anniversary of powered flight. Since its inception in 1947, the Air Force has proudly and vigilantly harnessed airpower in defense of freedom. From its earliest humanitarian operations during the Berlin Airlift to its ongoing operations in the war on terrorism, the United States Air Force, along with our sister services, have always answered the call to safeguard our nation's liberties and to deter and defeat aggression where ever it exists.

Sustaining this vital mission requires the dedication and commitment of every airman, but equally important, it requires the backing of the American people. Such public support hinges upon an informed and knowledgeable population. The United States Air Force Museum helps the service share its compelling story with the American public. Through its impressive aircraft collection, sensory-rich exhibits, and informative and educational events and programs, the museum supports all airmen by informing the public of the Air Force's history, mission and evolving capabilities. As the world's largest and oldest military aviation museum, the institution attracts nearly 1.3 million visitors a year, serving as a window on the Air Force.

As our nation confronts a new and dynamic security environment, it is more important than ever for Americans to fully understand the cost of freedom. The museum's remarkable collection and exhibits remind us that those entrusted with protecting America's liberty must maintain a commitment to vision and innovation to remain relevant in today's technology-driven world. The museum showcases examples of the honor, courage, and sacrifice that serve as the calling card for airmen of all generations.

The USAF Museum's legacy of excellence continues through the completion of a third, 200,000 square foot hangar scheduled to open in Spring 2003, coinciding with the year's Centennial of Flight celebration and a variety of major events in Dayton, Ohio, home of the Wright brothers. As the centerpiece of a multi-phase expansion program that includes a hall of missiles, space gallery and education center, the hangar will detail the story of the Cold War, emphasizing the Air Force's role in strategic deterrence, Soviet containment, and advanced technological development.

On behalf of the men and women of your Air Force, I encourage you to experience the United States Air Force Museum firsthand to see what makes ours the world's premier air and space power.

JOHN P. JUMPER
General, USAF
Chief of Staff

★ Contents

Foreword

When I was a boy, from time to time my father would take me to the Air Force Museum which way back then was located in an old WW II engine shop at Patterson Field. On one occasion, maybe 3rd grade, my class visited on a field trip. By today's standard this facility was tiny and crude. Many of the airplanes were housed outside as some still are today and, of course, the collection was much smaller. It didn't matter to a boy of 9 or 10. This place was special.

Of course, as a kid I could see all I needed to see in about an hour. I could hardly wait to get through the early aviation galleries. It seemed like my father paused there to read everything twice. WW II was definitely more interesting. The fighter aircraft of my father's generation were sleek and powerful, and everything about them said speed. The most memorable exhibit for me, however, was a cutaway B-52 cockpit with a loop movie running on the outside of the windshield that allowed a boy for a few minutes--until it was someone else's turn--to become a cold warrior at the controls of what is still today one of our most potent weapons systems, as well as a classic. I doubt much has changed with kids today. They still want to see the latest, biggest, highest flying, sleekest, and fastest the Air Force has to offer. They can also experience a much more sophisticated simulator that dives, banks, climbs, and rolls like the real thing.

I have just turned 45 and it was only in recent years that I began to fully appreciate the genius and superb craftsmanship of what is undoubtedly the museum's crown jewel. It is, of course, the 1909 Wright Military Flyer: a replica of the world's first heavier-than-air flying machine purchased by the military. It is where the story of the Air Force begins even though it was an Army machine. Every other aircraft you see in this great facility, even the SR-71, is its descendant. You might have had to walk by the Flyer at a fast clip if you brought children or grandchildren; or maybe you wanted to see WW II aircraft or the jets like I did as a boy. It's okay; you probably had a good excuse.

My own daughter and son do the exact same thing today, knowing full well (as their father did at their age) that their last name is Wright and their great, great granduncles were Wilbur and Orville Wright, the two gifted men who designed and built the original version of that 1909 machine, as well as the world's first successful heavier-than-air, powered flying machine in 1903. As I said before, kids haven't changed. But I eventually came around to appreciate Uncle Wil and Uncle Orv's beautiful creation as will my kids when they are 45.

The Air Force Museum is a living repository of so much more than just airplanes and spacecraft. It is an art gallery; a glimpse of what life was and is like in today's Air Force; a memorial to those who took off never to return; and, as a whole, a tribute to ingenuity, dedication to service, heroism, and creative genius sometimes born of curiosity, often born of necessity. I also hope you enjoy your visit to the city that is The Birthplace of Aviation: Dayton, Ohio.

Stephen Wright
Great Grandnephew of Wilbur & Orville Wright

★ 1 | The Growth of the Air Force Museum

"There shall be wings....If this accomplishment be not for me, 'tis for some other..." These words were spoken some 500 years ago by Leonardo da Vinci. Today we have wings and they are displayed for the public at the United States Air Force Museum, internationally recognized as the world's oldest and largest military aviation museum. It is located appropriately near Dayton, Ohio, at Wright-Patterson Air Force Base where the Wright brothers developed the first airplane and where they perfected their flying skills.

The Air Force Museum story is a moving one, an epic of men and machines, starting with ancient history and continuing through today—with the relatively recent inclusion of women in aviation. Our first thoughts about flight, Leonardo's experiments with helicopters, kites and gliders, ballooning, and dirigibles are all displayed for visitors as they first enter the museum's exhibit area. They quickly meet the Wright brothers and continue with the flying crates of World War I; the exciting "first" flights of the 1920s and 1930s; the action of World War II; and then continue through the ever-changing history of the Air Force as they proceed through the main museum complex. If they have allotted enough time, visitors may view more aircraft across the field in two WW II hangars.

As a national museum of aviation history, the Air Force Museum does not glorify war. It illustrates through historic airplanes, personalities, life-size dioramas, and other authentic displays the development of aviation and the development of airpower, primarily as embodied in the United States Air Force. And the museum points out how military aviation has brought major advancements to civil aviation. Knowledge gained from the four years of the First World War played an important role in developing civil aviation for the next twenty years. The same was true of later years. For example, the Boeing 707 jet transport, the early backbone of the commercial airline jet fleet, was developed from the Air Force's KC-135 tanker.

The Air Force Museum story is not static as exhibits constantly are being added or enlarged or moved, both indoors and out. Approximately 6,000 of the 68,000 items in the museum's collection are currently on public view. This includes more than 300 aircraft and major missiles on display. (Nearly 26,000 items are on loan to some 360 military and civilian museums and air parks around the world.)

Historic aircraft and personalities comprise the major displays, but they also include such things as clothing and diaries of flying aces, prisoner-of-war exhibits, Stumpy the homing pigeon, a German buzz bomb, an American chapel display, aviation art, capsules from early space exploration, and souvenirs from peace-keeping missions around the globe. Also, aircraft from Canada, France, Germany, Great Britain, Italy, Japan, and the Soviet Union are exhibited. The museum is a virtual paradise for the aviation buff or the camera bug. More important, it is a place of interest for the entire family, as attested to by the attendance of over one million visitors annually. Some tour the museum for two or three hours, others inspect the museum for two days and complain they did not allot enough time. Many return for repeat visits, even from foreign countries.

Former Air Force crewmen who visit the museum often stand in silent tribute before the representative aircraft that carried them around the United States, over foreign lands, and then safely home. Oftentimes they can be overheard relating their experiences to friends, their children, or grandchildren. Before they walk away, they nearly always reach over and tenderly touch the wing or propeller of the type airplane they flew years before. That gesture is like a final farewell to a faithful and trusted comrade-in-arms.

In addition to the aircraft and other exhibits, famous people in world aviation history are depicted at the Air Force Museum. The visitor can find displays honoring such people as Octave Chanute, who built gliders as a first step toward a powered flying machine; Capt. "Eddie" Richenbacker, the top American WW I ace; and the Tuskegee Airmen, who conquered two types of enemies in WW II. Unexpected, perhaps, is the exhibit honoring musician Glenn Miller who, as a major, commanded the famous Army Air Forces Band during WW II. However, the Glenn Miller band was acclaimed as "the greatest morale builder" in Europe, second only to letters from home. The museum tells the Glenn Miller story in a free-standing display which includes his original trombone (used by actor and WW II pilot James Stewart in the Glenn Miller movie), his "fifty mission crush" hat, eyeglasses, music case, and video story with Miller music.

From its meager beginnings in 1923 in a corner of a hangar, the Air Force Museum has expanded into a $25 million facility constructed in 1970-71 and enlarged in 1975-76, in 1985-87, and again in 1989-91. Ground was broken in 2001 for the construction of the first of four more buildings at a cost of $35 million. The main museum complex lies in the northwestern portion of the old Wright Field, and within three miles of Huffman Prairie where the Wright brothers developed the art of controlled flight in 1904 and 1905. There they later established a flying school and trained many of America's pioneer pilots. Today the area is a portion of Patterson Field which with Wright Field comprise Wright-Patterson Air Force Base. Wright Field is no less historic in its own right. Since its dedication in 1927, Wright Field has been known as an Air Force center for research and development. Numerous aircraft were tested and accepted here by the Air Force in the two decades before WW II.

Back in 1923, the first museum was located in Dayton at McCook Field, the Aeronautical Engineering Center established during WW I. The museum's mission called for the collection of technical intelligence related to American and foreign aircraft and equipment used in the previous war. It was moved to Wright Field in 1927 and occupied 1,500 square feet in a laboratory building. Then in 1935, the museum got its first real home, a $235,000 building provided by a federal works project. With a collection of about 2,000 items, the new museum at Wright Field was opened to the public in 1936. The two-story WPA (Works Projects Administration) structure is still visible from Springfield Pike near the railroad overpass about a mile northeast of the current museum.

With the advent of WW II, Wright Field was closed to the public and the museum building was converted to wartime office use. After the Allied victory over Germany and Japan, an engine overhaul building at Patterson Field was designated as the museum's new home. Curator Mark Sloan, who retired in 1972, began the search in 1946 for items for both the Air Force Museum and the National Air and Space Museum of the Smithsonian Institution. After eight years of acquiring and preparing items for display, the Air Force Museum opened in 1954 at its fourth location. But the 1954-71 museum was from the beginning only a temporary home. It was not fireproof or air-conditioned, and it had supporting posts every sixteen feet in one direction and every fifty feet in the other. All this made it unsuitable for properly displaying and protecting the museum's growing, priceless collection. By the early 1960s the building was outgrown, and there was little expansion room for the adjoining outdoor aircraft display area.

In 1960, three Daytonians chartered the Air Force Museum Foundation Inc. "to assist the United States Air Force Museum where federal funds are not available." They were banker C. Frank Scarborough, attorney James F. Barnhart, and radio disc jockey John Fraim. With the help of Dayton philanthropist Eugene W. Kettering, the foundation developed an active board of directors from representatives of industrial, governmental, educational, and philanthropic organizations. The board launched a fund drive that netted $6 million, largely from Kettering and members of his family; from local and national business and industry organizations, and private citizens; from members of the Air Force, including those in

Guests arriving for special evening activities at the U.S. Air Force Museum have had this view of the modern facility since May 1991. The 500-seat IMAX Theater is contained within the structure to the left while the five-story atrium lobby is to the right. *U.S. Air Force photo.*

active duty, reserve, and guard units; and from veterans. All this enabled construction of the new museum to begin in April 1970. The fifth and permanent museum was opened to the public in August 1971 with President Nixon dedicating it the next month. But Kettering, who had meant so much to the funding of museum, did not see the project completed; he died in 1969.

This permanent museum structure is 766 feet long and follows the design of two aviation hangars joined by a "core" administrative section. In addition to offices, it includes a 500-seat theater and two smaller galleries. Each of the hangars, or large galleries, has an arched roof that reaches eighty feet above the ground. The interior width is 240 feet, exactly twice the distance flown by Orville Wright on December 17, 1903. More than 230,000 square feet of space is incorporated in the structure, which includes 160,000 square feet of aircraft display space.

Approximately 100 aircraft and missiles are featured inside the 1971 structure. Early aviation history, Wright aircraft, those flown up to the start of WW II, and associated memorabilia are displayed in the smaller of the two hangars, known as the Early Years Gallery. The other hangar is known as the Air Power Gallery and is now devoted to the aircraft, personalities, and stories of the American victory over Nazi Germany and Imperial Japan in WW II.

The B-36 Peacemaker bomber was the first aircraft placed inside the Air Power Gallery. With a wingspan of 230 feet, it had to be moved into position before the hangar was completed. Its wing tips stretched almost from wall to wall, nearly twice the distance of man's first powered, controlled, and sustained flight. Like the first Wright airplane, the B-36 also has propellers mounted behind the wing. Then in late 2002, the B-36 was laboriously moved outdoors and towards its future home in the Eugene W. Kettering Building.

In the core area of the museum now also can be found a number of smaller displays, including the valuable Kettering collection of scale model planes, a portion of the Air Force Art collection, cartoon strips by Daytonian Milt Caniff, rotating exhibits, and displays honoring enlisted personnel. The entire museum structure is air-conditioned and heated to provide visitor comfort and to help preserve the older fabrics, wooden aircraft, and related artifacts.

The Carney Auditorium in the core area periodically offers free documentary films during the day throughout the year and evening guest lectures by aviation notables September through May, skipping December. The Kettering Hall along the north side of the auditorium often displays special exhibitions of aviation art. A permanent feature there is the 61-by-71-foot First Flight mural that reproduces the famous 1903 photograph through an intricate pattern of 163,296 one-inch ceramic tiles. Originally the mural was displayed in the Dayton Convention Center, but had been in storage for a decade until Kettering's widow, Virginia, paid to have it relocated at the Air Force Museum in August 1998. The exhibit hall on the south side of the auditorium featured walk-through dioramas on the fiftieth anniversary of the independent Air Force in 1997-98, of the Berlin Airlift in 1998-99, and of the Korean War in 2000-2002.

During 1975-76, a two-story addition was constructed in front of the core building for slightly more than $900,000 with funds provided by the Air Force Museum Foundation and the estate of Brig. Gen. Erik H. Nelson, who in 1924 participated in the first flight around the world. That project provided for more office space for the museum and its foundation, an enlarged cafeteria (with views of the outdoor attractions), and an expanded gift shop with bookstore, clothing, souvenirs, and model shop.

Then in 1977, to help preserve the many aircraft that had been displayed in the Outdoor Air Park since 1971, the museum moved most of them across the field to two recently vacated and adjoining WW II structures, Hangars 1 and 9. This temporary annex is now known as the Presidential Aircraft Hangar and the Research and Development/Flight Test Hangar. More than fifty aircraft and major missiles are displayed in Hangars 1 and 9. Current information is available in the museum lobby on how to visit there.

Ground was broken for another addition, the Modern Flight Hangar, in October 1985. It was completed in December 1987 and officially opened in April 1988. This continuous hangar-like structure measures nearly 800 feet by 200 feet, and is located behind and parallel to the first museum building. Half of the $10.8 million cost was voted by the Congress while the remainder was raised by public donations through the foundation. More than seventy aircraft and major missiles were displayed there, ranging from World War II to those recently introduced to Air Force flightlines. With Hangars 1 and 9, 10.5 acres of exhibits then were under roof, making the U.S. Air Force Museum the largest in the world.

A much different addition was completed and dedicated in May 1991, a giant wide-screen theater. Work on this 500-seat IMAX Theater began shortly after the contract was awarded in October 1989 by the Air Force Museum Foundation. The $7.3 million project also includes an 80-foot-high glass atrium over a new lobby which serves as the architectural focal point of the entire museum complex. Aviation and space movies are featured in the IMAX Theater. Its six-story screen, sophisticated sound, and tiered continental seating tend to bring viewers into the action projected in front of them. An admission fee is charged by the foundation, which operates the theater.

Proceeds from the theater, gift shop, and other foundation activities will continue to help pay for expansions at the museum. Ground was broken in mid-2001 for the Eugene W. Kettering Building which provides more than 200,000 square feet of additional exhibit space to present the story of the Cold War, the longest conflict in U.S. history. Also scheduled to follow are a Hall of Missiles in an above-ground silo design; a Space Gallery to present the Air Force's past and future roles in space; and a greatly expanded Education Center, particularly to benefit further

When this photograph was taken in the mid-1900s, more than 10-1/2 acres of aircraft and other exhibits were displayed under roof at the U.S. Air Force Museum. Earlier, in 1991, its IMAX Theater and atrium lobby (*left*) had been added by the Air Force Museum Foundation Inc. In June 2001, a ceremonial ground-breaking ceremony was held for the first of four additions to be built to the right. The first, the Eugene W. Kettering Building, would add 4-1/2 acres as the museum continued "Reaching New Heights." *U.S. Air Force photo.*

the youth of the nation. As the U.S. Air Force Museum continues "Reaching New Heights" with the support of the Air Force Museum Foundation, many more items will be placed on display for the first time, or be returned to public viewing. Among the latter are the intercontinental missiles that had been in storage since 1997 awaiting restoration and their new silo home, for which ground was broken in October 2002.

The museum's staff also plans for the acquisition of additional aircraft of historical importance. For instance, the museum had a long-standing request in with the federal government for the VC-137C known as SAM 26000. This aircraft has served every president from Kennedy to Clinton, and carried Kennedy's body home from Dallas. Finally, the aircraft made its last flight, landing at the museum on May 20, 1998. Most of the museum's aircraft, of course, are ones that have become surplus to Air Force needs. Some have come through loans or trades with private individuals or other museums. Occasionally a museum friend will purchase a particular airplane and donate it to the Air Force Museum. A few have come directly and indirectly from foreign nations. Others have been retrieved from remote locations, built from original plans, or restored with authentic parts. A video exhibit near the museum's O-38 tells how this observation airplane was retrieved and restored after lying in the Alaskan wilderness from 1941 to 1968.

Nearly a year before the current museum was opened in 1971, more than thirty aircraft were towed the seven miles from the former museum at Patterson Field to Wright Field. Others were moved later. Highway signs and signal lights had to be removed in some areas. The Air Force's only remaining XB-70 experimental bomber was one of those to make the circuitous trip. Workers of the Air Force Logistics (now Materiel) Command had to strip the supersonic jet of her six engines and ten tons of equipment to reduce her weight to the stress limit of a highway bridge along the route. The engines have since been reinstalled.

When President Nixon flew to Wright-Patterson AFB to dedicate the new facility officially on September 3, 1971, he told of Orville Wright's taking his father on a flight over Huffman Prairie and of the elder Wright's speaking these words: "Higher, higher, higher." The President continued: "That was the spirit of American aviation. That is the spirit of the American Air Force....And let that spirit, higher and higher, always be the spirit of the United States of America." Those words were spoken in the Air Power Gallery against the backdrop of the B-36 Peacemaker and before a crowd of 12,000 spectators, including those hearing the address via loudspeakers outdoors.

Although the museum is owned, operated, and maintained by the Air Force, the private Air Force Museum Foundation continues its support on a daily basis. Through its offices at the museum, the foundation operates the museum gift shop, IMAX Theater, and Morphis MovieRide Theater; contracts for the operation of the cafeteria; raises funds; and generally assists wherever and whenever possible and practicable. When government resources were not available, for example, the foundation purchased wheelchairs and picnic tables for visitors and obtained a special display case and protection system for the museum's moon rock. Contributions to the foundation are tax deductible. Since 1978 the foundation also has operated a Friends of the Museum membership program to draw together aviation enthusiasts who are interested in the activities of the museum and in furthering its aims. For more information on the foundation, call (937) 258-1218 or write the Air Force Museum Foundation Inc., PO Box 33624, Wright-Patterson AFB, OH 45433. Or visit the foundation's web site at http://www.afmuseum.com at your convenience.

The museum also receives a great deal of daily support from some 400 volunteers who assist in virtually every aspect of its mission, from greeting visitors to restoring aircraft engines to helping in offices. They include homemakers, veterans, retirees, and persons still employed. A dedicated group of volunteers from the base's Officers' Wives' Club launched the volunteer program in 1972 with tours for school children. The museum's Education Division has since expanded the youth program to include a variety of special tours and programs for school and youth groups, families, and teachers. Most activities require advance reservations.

One of the objectives of the national museum of the Air Force is to serve as an educational tool to teach young people about aviation. The Secretary of the Air Force in 1971, Robert C. Seamans, Jr., perhaps best summarized the goal of the museum in these words: "The new Air Force Museum will serve as a tribute to all Americans who have contributed so much to the field of aviation. It will also serve as an inspiration to future generations of Americans to increase their knowledge and awareness of the United States Air Force and the history of flight." As Ohio's top free attraction since his remarks were made, the U.S. Air Force Museum continues to fulfill its historical preservation and educational missions with international acclaim.

A different perspective about aviation is offered to visitors of all ages who tour the Memorial Park on the south side of the main museum buildings. There are located some 450 memorials that have been dedicated to commemorate the military service of individuals and of organizations. Whether it be a tree with a simple plaque or an elaborate granite monument, each memorial was financed and dedicated by interested citizens. Then Vice-President George H.W. Bush participated in one such dedication.

As the twentieth century drew to a close, the National Aviation Hall of Fame was relocated from downtown Dayton to its own new building on the museum grounds. It was constructed between the northern portions of the Modern Flight Hangar and the Air Power Gallery, covering more than 17,000-square feet. Founded in 1962, the hall has enshrined more than 178 national aviation heroes, military and civilian. Several of the exhibits in its Learning and Research Center focus on enshrinees connected with aircraft displayed at the museum. The National Aviation Hall of Fame and the United States Air Force Museum, it should be noted, are distinct and separate organizations.

Located six miles northeast of Dayton, the Air Force Museum is open to the public without parking or admission charge seven days a week from 9 a.m. to 5 p.m. There is a charge for the IMAX Theater and for the Morphis MovieRide Theater. The museum is closed only on Thanksgiving, Christmas, and New Year's Day. Interstate Highways 70 and 75 conveniently guide motorists to the Dayton area; Ohio Route 4, Needmore Road, Interstate 675, U.S. Route 35, and Woodman Drive further route motorists to the museum. (Woodman Drive, Harshman Road, and Needmore Road form the Wright Brothers Parkway, another access to the museum area.) Photography is encouraged at the museum with flash and high-speed film recommended indoors. Facilities are available for persons with disabilities. Pets are not permitted. Other areas of Wright-Patterson AFB, excluding Hangars 1 and 9, are restricted and closed to the public.

For current information, call (937) 255-3286 or write the U.S. Air Force Museum, 1100 Spaatz St., Wright-Patterson AFB, OH 45433-7102. Those with internet access may visit the museum's award-winning web site at http://www.wpafb.af.mil/museum/ for a virtual tour of the museum and up-to-date information on special events and new attractions. The web site has received as many as 1.2 million hits in one month and has been accessed from more than fifty countries.

Visitors arriving at the Air Force Museum in the early 1970s had this view. President Truman's **Independence** was visible between the Atlas and the shorter Thor missile. The indented core area of the museum was extended forward in 1975-76 and again in 1989-91, with the missiles being removed in 1997.

Patriotic bunting decorated the museum on September 3, 1971, when the new $6 million facility was dedicated by President Nixon. It also rained that day, but twelve thousand people braved the weather to attend the ceremony. *U.S. Air Force photo.*

UNITED STATES AIR FORCE MUSEUM

SINCE THE WRIGHT BROTHERS' FIRST FLIGHT, THE HISTORY OF AVIATION HAS BEEN MARKED BY THE COURAGE AND CONSUMING DESIRE FOR KNOWLEDGE WHICH HAVE ENABLED MANKIND IN A RELATIVELY FEW YEARS TO LEAVE THE EARTH AND BEGIN THE EXPLORATION OF SPACE.

THIS MUSEUM WHICH IS MEANT TO PRESERVE THE HERITAGE OF MILITARY AVIATION, WAS BUILT BY THE AIR FORCE MUSEUM FOUNDATION UNDER THE LEADERSHIP OF EUGENE W. KETTERING AND WITH THE GENEROUS VOLUNTARY SUPPORT OF U.S. AIR FORCE PERSONNEL, INDUSTRY, AND INTERESTED INDIVIDUALS.

ON BEHALF OF A GRATEFUL NATION, I JOIN IN DEDICATING THIS AIR FORCE MUSEUM TO THE AMERICAN PEOPLE AND TO THE THOUSANDS OF AIRMEN WHOSE PIONEERING SPIRIT, DEVOTION TO DUTY AND LOVE OF FLYING HAVE CONTRIBUTED SO MUCH TO OUR PROGRESS IN THE AIR AND TO OUR HAPPINESS ON THE EARTH.

RICHARD NIXON

SEPTEMBER 3, 1971

This plaque hangs on the outside wall near the museum entrance.

General Jack G. Merrell, commander of the Air Force Logistics Command, opened the official ceremonies with President Richard Nixon seated directly in front of the B-36 Peacemaker. Key platform guests included Mr. Frank G. Anger, president of the Air Force Museum Foundation; Ohio Governor John J. Gilligan; and Mr. Robert S. Oelman, chairman of the board, Air Force Museum Foundation. *U.S. Air Force photo.*

President Nixon accepted the museum from the Air Force Museum Foundation, Inc., on behalf of the American people. He called on the nation to continue leading the world in aviation. *U.S. Air Force photos.*

Museum dedication guests had an opportunity to view the older airplanes in the Early Years Gallery which has since been partitioned into a maze that directs visitors on a chronological walk through aviation history. *U.S. Air Force photo.*

Secretary of the Air Force Robert C. Seamans, Jr., presented a model of the museum to Mrs. Eugene W. Kettering, whose late husband spearheaded the campaign for the new facility. Gen. and Mrs. Jack G. Merrell look on. The general commanded the Air Force Logistics (now Materiel) Command which provides much support for the museum. *U.S. Air Force photo.*

This metal plaque is a full-scale reproduction of the *Dayton Daily News* special edition which was distributed at Wright-Patterson Air Force Base when the museum was dedicated by President Nixon. It includes an article referring to Dayton, Ohio, as the birthplace of aviation. It also includes a reference to a future era of unmanned military aircraft called "remotely piloted vehicles."

The growth of the Air Force Museum from its meager beginnings in a corner of a hangar in 1923 until today is shown in this display, created for its seventy-fifth anniversary. In the bottom right is an artist's early concept for the expansion of the museum.

A corner of this McCook Field hangar housed the first museum from 1923 until 1927. Originally the museum was called the Engineering Division Museum. *U.S. Air Force photo.*

Some of the World War I airplanes displayed at the McCook Field museum. Wings were removed so that more planes could be housed in the limited space available. *U.S. Air Force photo.*

In 1927 the museum's artifacts were moved to the new Wright Field. Since no museum building was available there, engines, propellers, other items of aeronautical equipment, and a few airplanes were exhibited in the back of this laboratory building until another move in 1935. In 1932 the collection was renamed the Army Aeronautical Museum. *U.S. Air Force photo.*

A special museum building was constructed at Wright Field, and in June 1935 the museum moved into it. Unfortunately, the world situation in 1940 forced the closing of the museum so the building could be used for offices by the rapidly expanding United States Army Air Corps. Its artifacts were stored wherever space was available for the duration of World War II. *U.S. Air Force photo.*

An interior view of the 1935 building showing some museum displays. A need for an aviation museum in the United States had been long recognized. Pioneers professed that many hours of needless engineering research might be saved if examples of what had already been accomplished were readily available. *U.S. Air Force photo.*

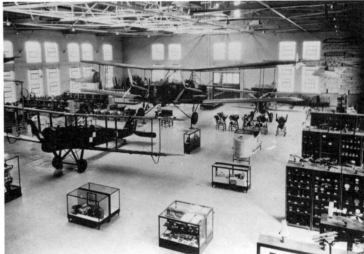

Following World War II an engine overhaul building at Patterson Field was obtained for the museum, and on January 2, 1948, the Air Force Technical Museum was officially established as the successor to the Army Aeronautical Museum. Originally the Technical Museum displayed only technical items such as engines and cameras, but in 1954 it began to acquire full-size airplanes for public exhibit. Two years later it officially became the Air Force Museum. *U.S. Air Force photo.*

The fourth museum looked like this in its final years. During the Christmas holidays the Atlas missile was decorated to resemble a giant Santa Claus. Many of the airplanes which were formerly parked outdoors, including the giant B-36, are now displayed indoors. *U.S. Air Force photos.*

Since World War II, the following have served as directors of the Air Force Museum:

Mr. Mark Sloan	1946-56
Col. John F. Wadman	1957-58
Maj. Robert L. Bryant, Jr.	1958-62
Lt. Col. Kimbrough S. Brown	1962-63
Capt. Burdette E. Townsend	1963-64
Col. William F. Curry	1964-67
Col. Joseph D. Hornsby	1967-70
Col. Bernie S. Bass	1970-76
Col. Richard L. Uppstrom*	1976-96
Maj. Gen. (Ret.) Charles D. Metcalf	1996-

The following have served as executive secretaries of the Air Force Museum Foundation Incorporated:

Wilma Donley Hatcher	1965-71
Lt. Col. (Ret.) Frederick C. Wolf	1972-89
Col. (Ret.) Richard A. Johnson	1989-

*Colonel Uppstrom served the last half of his directorship as a civilian.

Some of the items on exhibit during the 1950s. Numerous pillars and a relatively low ceiling in the building often interfered with properly displaying the exhibits. *U.S. Air Force photo.*

This B-25 bomber led a B-24 and a string of other aircraft along Route 444 from the old museum at Patterson Field, past Wright Brothers Hill, to their new home at Wright Field. Several governmental agencies cooperated in this unusual "flight." *U.S. Air Force photo.*

Wing tips had to be temporarily removed from this B-24D Liberator so that it could be towed to the new museum. Thirty airplanes were moved on two consecutive weekends. *U.S. Air Force photo.*

The direct route to the new museum would have been via the road to the left, but a narrow railroad bridge necessitated a longer route freer of obstructions. *U.S. Air Force photo.*

Before the XB-70 could be moved, ten tons of equipment and six jet engines had to be stripped from the Valkyrie so that it would not damage the bridge. Traffic signals had to be removed from some intersections to permit the aircraft to pass through. Wright Brothers Hill is in the background. *U.S. Air Force photo.*

After its seven-mile journey to the new museum, the once-secret XB-70 invited visitors to inspect it and the museum closely. Aircraft, such as the KC-135 tanker, regularly pass overhead at a low, but safe altitude. They provide an added dimension to the static airplanes on display. The XB-70 initially was moved indoors in 1987 after the second exhibit hangar was completed. In 2002 it was moved again, across the field to the Research and Development/Flight Test Hangar.

Six-inch-diameter pins fasten the parabolic museum roof to concrete piers which weigh three-and-a-half tons each. President Truman's VC-118 airplane is also pictured. It was moved in the late 1970s across the field to the Presidential Aircraft Hangar.

Contractors' plaque

This large plaque was a gift to the museum from the plasterers who worked on the initial building which opened in 1971. It was made on their own time and presented as a memento of their interest in the museum.

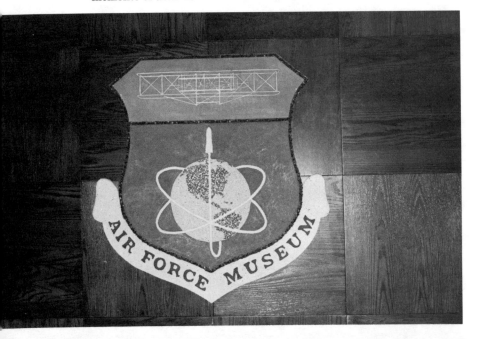

Already a place where a child's imagination can take flight, the museum also has participated in the "Read Across America" celebration honoring Dr. Seuss. His book, "Oh, the Places You'll Go," encourages readers to fly high and to strive for lofty goals. At locations throughout the museum, base volunteers aided the Education Division in supporting the national read-to-youth project. *U.S. Air Force photos.*

Hundreds of children and adults annually participate in the Kite Festival at the museum. In the process of making and flying their creations, children in particular learn the principles of flight while enjoying an outing with their families. Both amateurs and professionals participate in various aspects of the Kite Festivals. *Air Force Museum Foundation photo by Jerry Rep.*

Aviation-related souvenirs, gifts, and models of all sizes and descriptions are available at the gift shop operated by the Air Force Museum Foundation. The museum bookstore, located along a facing wall, is the largest aviation bookstore in any aviation museum in the United States. Profits from these shops and a 300-seat restaurant upstairs help the foundation purchase needed items which the museum is unable to obtain through government sources.

The foundation's Friends of the USAF Museum program also assists the museum through a membership program. Benefits from a Regular category annual fee include a membership card and certificate, discounts on purchases and orders, colorful aircraft calendar, a museum aircraft brochure, and quarterly Friends Journal. Higher category memberships, with increased benefits, are available for individuals, organizations, businesses, and corporations. Write: A.F. Museum Foundation Inc., PO Box 1903, Wright-Patterson AFB, OH 45433. Telephone: (937) 258-1225.

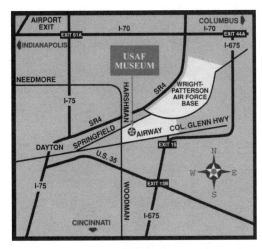

Main highways in Ohio lead visitors to the Air Force Museum. After entering the main parking lot, visitors soon see Valor Park (*right*), which honors the Air Force's 59 Medal of Honor recipients. Valor Park was dedicated in May 2002 with all six living recipients in attendance. *Air Force Museum Foundation photo by Jerry Rep.*

Some sixty aircraft are displayed in the Modern Flight Hangar, where they are grouped by types of missions flown. The Chaplain Service exhibit is shown between the Douglas C-124C Globemaster II and the Douglas B-18A Bolo. The central floor area is kept open for special museum events.

The North American XB-70 Valkyrie once dominated the northern end of the Modern Flight Hangar, as it now dominates the Research and Development/Flight Test Hangar. In the foreground are a Martin X-24B and a Douglas X-3 Stilleto.

The Growth of the Air Force Museum 19

Larger aircraft dominate the Outdoor Air Park, including the Lockheed EC-121D Constellation (*left*) and the Boeing NKC-135A airborne laser laboratory (*tail, right*). In the distance can be seen a likeness of the Wright brothers on the museum's Presidential Aircraft Hangar. The airplanes outdoors are periodically rotated to other areas of the museum.

For three decades the Convair B-36J Peacemaker dominated the Air Power Gallery (*right*). Then in October 2002, as construction progressed on the Eugene W. Kettering Building (*bottom photo, left rear*), numerous display panels, many aircraft, and large sections of the hangar's northeast end were removed in order to tow the B-36J towards its new home in the new gallery.

★ 2 | *Early Aviation History*

Man's early concerns about aerial flight and our first attempts to slip the bonds of earth are portrayed at the Air Force Museum in models, sketches, photographs, words, and a few select airplanes. In this prologue to flight are presented the basic origins of our eventual conquest of air and space. The history of flight begins with a representative of a bas-relief stone sculpture of Ashur, the winged deity who was venerated by the Assyrians hundreds of years before Christ. The history progresses quickly and chronologically to the world's first powered, sustained, and controlled heavier-than-air flight on December 17, 1903, at Kitty Hawk, North Carolina. Few people of the day recognized the revolutionary implications of the flight by Orville Wright of Dayton, Ohio. The world as they knew it had thus been reduced in scope.

Understandably, the Air Force Museum devotes significant attention to the Wright brothers and their accomplishments. A few scraps of wood and fabric from the original 1903 Flyer are displayed, as well as original propellers and a bicycle from the Wright Cycle Company. More important, there is also a full-scale reproduction of the Wright 1909 Military Flyer (the world's first military airplane) and an original Wright 1911 Modified "B" Flyer.

Displays on the walls around these two air machines illustrate achievements in the development of aircraft by both the Wrights and the aviation wing of the Army. The latter had been established on August 1, 1907, by the War Department as the Aeronautical Division of the Office of the Chief Signal Officer of the Army. Command of what was to become the United States Air Force was given to Charles deForrest Chandler, who was described as "a dashing young captain with a venturesome spirit." He was ordered to take charge of "all matters pertaining to military ballooning, air machines, and all kindred subjects." Two enlisted men and a civilian clerk were assigned to his staff.

In 1910, the Aeronautical Division had grown to fourteen men, including three flying officers. Two of them were later reassigned, so it fell to Lt. Benjamin Foulois to accept the first military airplane from the Wright brothers. Along with this responsibility, he was told to "assemble this flying machine and teach yourself to fly." With the aid of correspondence lessons from the Wright brothers, he did. One year later, Foulois and Chandler were joined by Lt. Henry H. "Hap" Arnold, who was to become the first General of the Air Force.

Early flyers, military and civilian, seemed bound together in a fraternity accepting a common challenge. Leaving the comfortable earth for the alien element of the skies demanded a blend of courage, curiosity, dedication, and daring. They played a dangerous game and accepted the consequences of losing. When Orville Wright conducted tests for the War Department in September 1908, Lt. Thomas E. Selfridge was his passenger. In the crash that ended these tests, Lieutenant Selfridge was fatally injured. He thus became the first to die as the result of an airplane accident. He also had been the first military man to fly an airplane alone. As fate would have it, he paid the price that airmen have faced since they first ventured off the ground.

In spite of dynamic personalities and leadership, however, the nation's love affair at the time was with the automobile and not with the airplane. The nation that gave birth to the airplane was following a course that would have us entering World War I with a twelfth-rate air arm. As WW I developed in Europe, the airplane was still considered a toy by many in America. A few realists, struggling against skeptics and inertia, developed the fledgling air force as best they could. These efforts, too, are presented to museum visitors as then wend their way through the exhibition maze in the Early Years Gallery to the end of the first major section of the museum.

Man's earliest dreams of flight welcome visitors to the Air Force Museum. Khensu, the upright creature with four wings, was the Egyptian Navigator of the Skies about 1000 B.C. To the left of the large sign is Ashur (not seen), the Assyrian winged deity.

Ashur and Khensu.

This prologue-to-flight section stirs the imagination of museum visitors. Flights attributed to mythology and actual flights in balloons and strange contraptions led to the Wright brothers' breathtaking flight at Kitty Hawk.

To achieve powered flight man first had to solve the problems of control and aerodynamic lift while developing a satisfactory lightweight engine. It required a hundred years of effort by many air-minded men before the Wright brothers first demonstrated powered, controlled, and sustained flight.

A bronze sculpture of Daedalus and his son Icarus hangs as a mute reminder that even the early Greeks were interested in flight. According to the myth, Daedalus and his son Icarus fashioned wings of wax and feathers. Icarus disobeyed his father's warning, flew too close to the sun, and fell to his death. The sculpture depicts Daedalus bearing up his dead son.

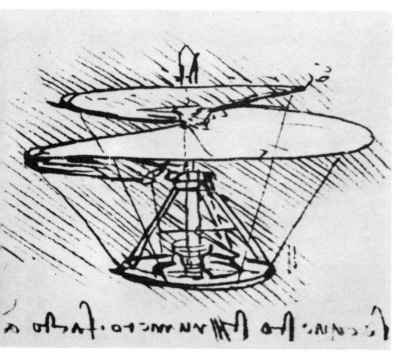

A very early pioneer in the study of flight was Leonardo da Vinci, who made this helicopter drawing in about 1500. Ridiculed about his concern with flight, Leonardo took to writing in code: the left-handed artist wrote backwards.

Leonardo's early helicopter drawing has been transformed by museum staff members into an animated model—one of the more popular displays with the younger visitors. Push a button and the contraption revolves within its plexiglas case.

The Montgolfier balloon lifted man in his first aerial flight on November 21, 1783, when Jean-François Pilâtre de Rozier and the Marquis d'Arlandes went aloft for a flight which lasted approximately twenty-five minutes and covered more than five miles across Paris. The two daring fliers alternately fed straw and wool to the fire which was suspended in a grate beneath the open end of the balloon. They watched closely to see that sparks did not ignite the flimsy linen-and-paper balloon overhead.

Military ballooning in the United States began early in the Civil War. Best known of the aeronauts was Thaddeus S. C. Lowe. He and others made numerous aerial observations during the first two years of the war. At one time there were seven balloons in service with the Union Army. The South had at least three balloons in service. Both sides suspended balloon operations in 1863.

Correspondence from Abraham Lincoln and a telegram concerning ballooning during the Civil War period. The telegram is signed T. S. C. Lowe, "Chief Aeronaut, U.S.A."

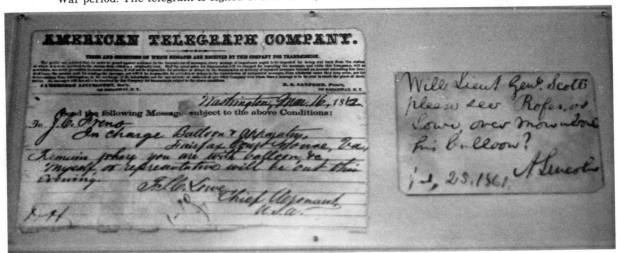

After the Civil War, there was no further military ballooning in the U.S. until 1892 when a balloon section was established within the Signal Corps. When war broke out in 1898, the Army's "air arm" consisted of one hand-sewn balloon. Despite incredible difficulties, Lt. Col. Joseph Maxfield succeeded in getting the balloon to Cuba where several ascents were made, including one in preparation for the famous charge up San Juan Hill. In 1899, the balloon detachment was disbanded and military aeronautics faded until 1907.

From the beginning, the usefulness of the balloon depended upon giving it "dirigibility" or directional control. This first demonstration flight of a Zeppelin occurred July 2, 1900.

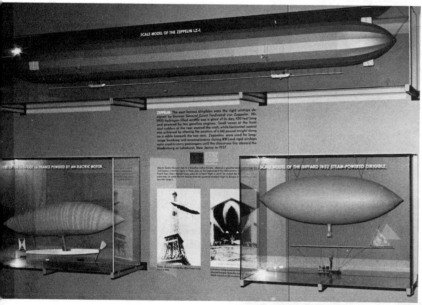

Three scale-model dirigibles are pictured in this display. At the top is the 420-foot German Zeppelin LZ-1, which flew in 1900. Left is the La France of 1884 and right the Giffard 1852 steam-powered dirigible.

Man's early attempts at flight included experiments with elaborate kites and wings.

Otto Lilienthal was the greatest of the glider experimenters. He built his first glider in 1891 and within the next five years this brilliant German made more than two thousand glides. He was fatally injured in a gliding accident in 1896.

Sir Hiram Maxim, an American-born inventor, tested his giant steam-powered flying machine in 1894 on a half-mile track fitted with guard rails to prevent the craft from rising more than a few inches. But the machine broke through the rails, and Maxim stopped the engines, as well as further flying. He noted: "Propulsion and lifting are solved problems; the rest is a mere matter of time." Maxim made no mention of the third requirement of successful flight, proper control.

Octave Chanute. A successful civil engineer, Chanute applied his knowledge of bridge building to the design of gliders. Published in 1894, his classic volume *Progress in Flying Machines* brought together in one book a history of man's attempts to fly. No less important to aviation history was Chanute's role as friend and advisor to the Wright Brothers. Above is a model of Chanute's 1896 biplane glider, his most successful design.

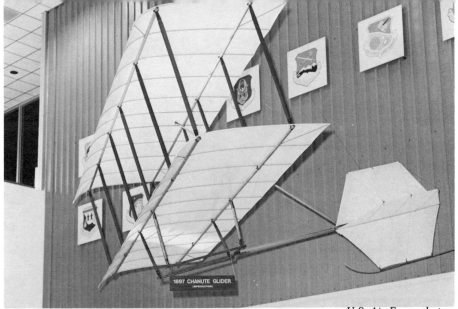

U.S. Air Force photo.

Charles Manly (left) and Samuel Pierpont Langley.

Langley's "aerodrome" prior to falling into the river.

Langley-type propeller.

SAMUEL PIERPONT LANGLEY

An astronomer and Secretary of the Smithsonian Institution, Langley in 1896 flew a steam-driven airplane model three-fourths of a mile and in 1898 received a Congressional grant of $50,000 for further development. Convinced now that a gasoline engine offered more promise than steam, he and his gifted assistant, Charles M. Manly, designed and built a revolutionary aircraft powered by a 125 lb. 53 hp. gasoline engine. On October 7, 1903, Manly attempted to fly from the deck of a houseboat on the Potomac River, but the airplane apparently fouled some portion of the catapult mechanism and tumbled into the river. Manly tried again on December 8 and again the attempt failed. These failures, plus the cruel jeers of the newspapers and cynics, crushed Langley's spirit and he retired. Nine days later, as the world continued to jeer, the Wrights flew on a bleak beach in North Carolina.

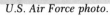

1897 CHANUTE GLIDER

Octave Chanute was already a successful American civil engineer in his 60s when he first became fascinated by aviation. In 1894, his studies led him to publish a compilation of early aviation knowledge and history. Progress in Flying Machines (republished in 1976). In addition to his significant contribution to aviation as a collector of aeronautical data, in 1895 he began designing and building gliders. These ranged from one using six tiers of wings to the successful biplane design which this reproduction represents. Mr. Chanute was aided by Mr. Augustus M. Herring, a civil and mechanical engineer, and Mr. William Avery, both of whom acted as pilot in various experimental and demonstration flights. Some of the gliders which Chanute designed and tested had either moveable wings or tail control surfaces; however each also relied on the pilot's body movements for proper equilibrium control. During the Wright Brothers' early experiments with gliders beginning in 1900, Chanute offered them encouragement and advice based on his own observations.

This reproduction represents Chanute's 1897 glider which was flown by Mr. William Avery 84 times at the St. Louis World's Fair in 1904. The craft was built in Chanute, Kansas by Mr. Johnny L. Litchenburg, using materials and construction methods similar to those employed in the fabrication of the original glider. Spruce and mahogany were used extensively in the original and the reproduction, as was brazing to join various metal components. One departure from the original techniques was the use of dacron rather than silk as the wing covering for greater resistance to deterioration through age while on display.

The glider was donated to the United States Air Force Museum by Octave Chanute's great-grandson, Mr. Octave A. Chanute, Director of the Historical Aircraft Research and Development Company of Denver, Col. It was jointly dedicated to the United States Air Force and the Chanute Technical Training Center in 1978.

After several experiments with steam-driven aircraft models, Samuel Pierpont Langley finally turned to the gasoline engine.

A wall display shows events leading up to the invention of the Wright brothers' gasoline-powered flying machine. Top right in the large display board is their Wright Cycle Company. Lower right is their first glider being flown as a kite. Closeups from the big board are included here. In 1901 and 1902 they developed and tested over fifty airfoil sections in a homemade wind tunnel and on a modified bicycle. In 1903 there was a photographer on hand to record their magnificent achievement.

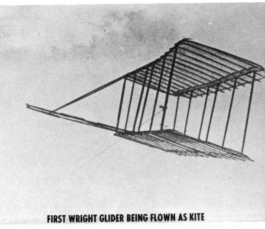

FIRST WRIGHT GLIDER BEING FLOWN AS KITE

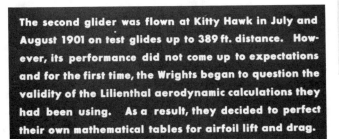

The second glider was flown at Kitty Hawk in July and August 1901 on test glides up to 389 ft. distance. However, its performance did not come up to expectations and for the first time, the Wrights began to question the validity of the Lilienthal aerodynamic calculations they had been using. As a result, they decided to perfect their own mathematical tables for airfoil lift and drag.

Third glider fitted with twin controllable rudders.

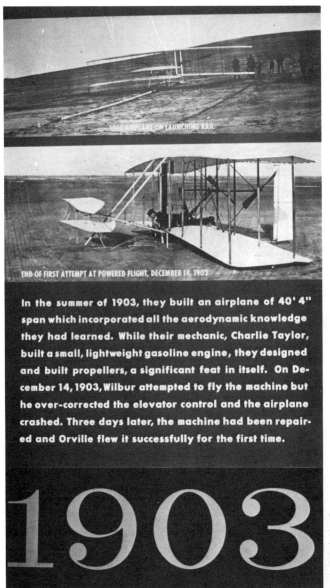

1903 AIRPLANE ON LAUNCHING RAIL

END OF FIRST ATTEMPT AT POWERED FLIGHT, DECEMBER 14, 1903

In the summer of 1903, they built an airplane of 40' 4" span which incorporated all the aerodynamic knowledge they had learned. While their mechanic, Charlie Taylor, built a small, lightweight gasoline engine, they designed and built propellers, a significant feat in itself. On December 14, 1903, Wilbur attempted to fly the machine but he over-corrected the elevator control and the airplane crashed. Three days later, the machine had been repaired and Orville flew it successfully for the first time.

1903

Take-off of the 1903 Wright Flyer on the world's first powered, sustained, and controlled heavier-than-air flight, December 17, 1903 at Kitty Hawk, North Carolina. Piloted by Orville Wright, the airplane remained aloft for 12 seconds and flew a distance of 120 feet.

The historic scene above was created from 163,296 one-inch ceramic tiles. This 61-by-17 foot mural originally was displayed in the Dayton Convention Center. Designed by Read Viemeister of Yellow Springs, Ohio, and built in Casalbuttano, Italy, in the early 1970s, it had been in storage for a decade until donated in 1998 to the museum by the National Aviation Hall of Fame. Mrs. Virginia Kettering paid to have it installed in the hall named after her late husband.

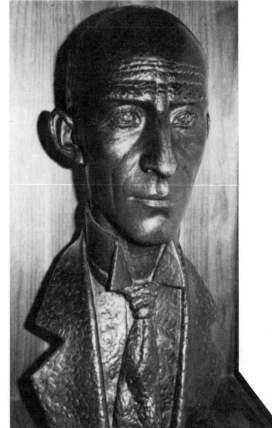

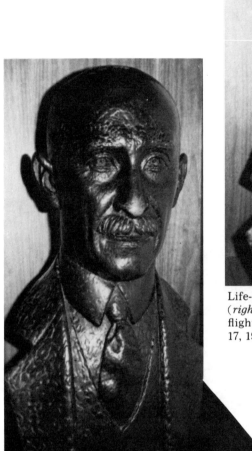

Life-size busts of Orville (*left*) and Wilbur Wright (*right*) flank the large photograph of their first flight at Kitty Hawk, North Carolina, on December 17, 1903.

An exact reproduction of the Wright 1909 Military
Flyer built by museum craftsmen, this plane dis-
plays an engine donated by Orville Wright. He is
depicted seated on the right. At his side is Lt. Frank
P. Lahm, one of the first trained to fly by Wilbur
Wright. Also shown are early aviators Lt. Frederic E.
Humphreys (*far left*) and Lt. Benjamin D. Foulois.
The latter learned to pilot the 1909 Flyer in 1910 via
correspondence with the Wrights. Overhead is a
Consolidated PT-1 trainer, first procured by the Air
Service in 1925.

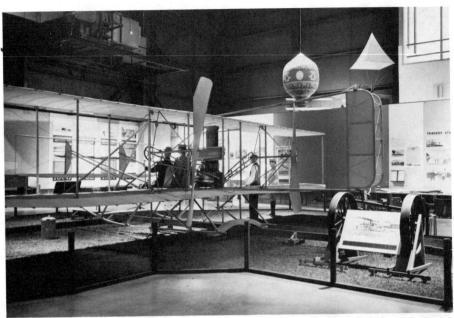

Rear view of the world's first military heavier-than-air flying machine, which had a maximum speed of forty-two miles per hour. The 1909 Military Flyer had skids similar to the X-15 which in 1966 set an unofficial world's speed record of 4,520 miles per hour. Maneuvering wheels for the 1909 Flyer are in the foreground, also seen here separately in closeup.

Maneuvering wheels were placed under 1909 Flyer to permit the aircraft to be moved while on the ground; they were detached prior to flight.

Resting unobtrusively behind the 1909 Military Flyer is an original Wright brothers bicycle, purchased from the Wright Cycle Company in 1895.

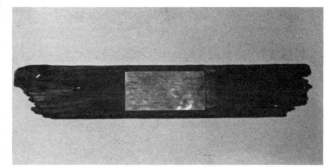

Piece of Wright brothers' hangar on Huffman Prairie given to Mr. E. N. Findley by Orville Wright in 1915. On September 20, 1904, Wilbur Wright made history's first full-circle airplane flight when he took off from Huffman Prairie and landed near the spot from which he had taken off.

During 1904 and 1905, the Wright brothers continued their research into the mysteries of flight over Huffman Prairie, now part of Wright-Patterson Air Force Base. On two occasions in 1905 they offered their invention to the U.S. Government. But their offers were rejected because few people actually believed they had invented a successful flying machine. This scene is at Simms Station, Huffman Prairie, November 16, 1904.

Orville Wright prepares to take off in the 1908 Flyer with Maj. George O. Squier as his passenger on September 12, 1908. The front of the Wright Flyer is to the right.

Orville Wright circles Fort Myer, Virginia, during tests conducted in 1908 for the War Department. During these tests, he made the world's first flight of longer than one hour. *U.S. Air Force photo.*

Signal Corps No. 1, a Wright Flyer, is readied on the launching rail at College Park, Maryland (1909). A weight dropped from the tower in the background pulled a rope that catapulted the aircraft along the rail and into the air.

An original Wright 1911 Modified "B" Flyer shares the attention of museum visitors with wall exhibits. The 1911 Flyer is the first plane produced and sold in quantity by the Wright brothers. Displays on the walls concern the first bombs dropped from airplanes, first weapons fired from flying machines, and early U.S. aviation schools at College Park, Md.; Texas City, Tex.; San Diego, Calif.; and the Philippines.

This Wright 1911 Modified "B" Flyer is one of the Wrights' original planes and was last flown in the 1924 International Air Race in Dayton. Overhead is the 1909-type Blériot which was used for both air and ground pilot training. It achieved fame when designer Louis Blériot of France flew it to England, making it the first plane flown across the English Channel.

An eight-cylinder Rausenberger engine powered this 1911 Flyer to a top speed of 45 mph. With the engine placed off center, the weight of the pilot and passenger was relied upon to help balance the aircraft.

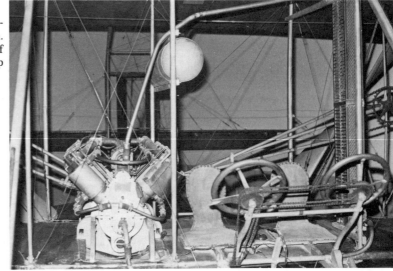

Shown are three wind tunnels, devices used for calibrating airspeed instruments and testing airfoils. The smallest tunnel is a replica of the one designed and built by the Wright brothers in 1901. It produced winds of from twenty-five to thirty-five miles per hour. This replica was constructed under the personal supervision of Orville Wright prior to World War II. The Orville Wright Wind Tunnel was designed by him in 1916 to conduct aerodynamic experiments during World War I at wind velocities greater than 160 miles per hour. The largest wind tunnel, seen here, was designed and built at McCook Field in Dayton in 1918. It has a twenty-four-blade fan sixty inches in diameter which achieved a maximum speed of 453 miles per hour at its fourteen-inch-diameter choke-throat test area.

The *Springfield* (Ohio) *Daily News* of May 30, 1912, ran an editorial drawing on the bottom of page one depicting an angel reaching out to receive Wilbur Wright on "His Last Flight." He had died of typhoid fever in Dayton. The newspaper also commemorated Memorial Day.

The *Dayton Daily News* of February 2, 1948, reported that "Civic life in Dayton came to virtual standstill Monday afternoon as Dayton paused to honor Orville Wright." A related headline called Orville a "Simple Man of Genius."

The first Gordon Bennett Balloon Trophy Race was won in 1906 by Lt. Frank P. Lahm and Maj. Henry B. Hersey when they traveled 402 miles across France in a free balloon. The first trophy, permanently awarded to Belgium in 1924, was donated in 1906 by James Gordon Bennett, an American newspaper publisher. The United States was awarded permanent possession of this second trophy in 1928 as a result of winning it for a third successive year when Capt. William E. Kepner and Lt. William Eareckson flew 460.9 miles in a free balloon from Detroit. It was donated by the Aero Club of Belgium.

Corporal Edward Ward, the first enlisted man to be assigned aviation duties in the Signal Corps, and other members of the balloon detachment, October, 1907. Early in 1907, the U.S. Army had become interested in ballooning and purchased two hydrogen balloons. One of these made a flight from Washington, D.C. to Harrisburg, Penna., on June 4, 1907. The Army observer on this flight was Capt. Charles deForest Chandler. Ballooning activities increased significantly during the following months as Army personnel gained aeronautical experience. Ward is in the center, seated.

During the 1910–1911 period, the Signal Corps had so few airplanes that it adopted a policy of granting its pilots necessary leave from duty to fly manufacturers' airplanes at civilian flying meets. At one such meet sponsored by the Aero Club of America on September 26, 1911, at the Nassau Boulevard Aerodrome on Long Island, Lt. Thomas Dewitt Milling, seen here, set a world endurance record of 1 hour 54 minutes 42.6 seconds with two passengers, for which he was awarded the Rodman Wanamaker Endurance Trophy. Milling, together with Lt. H. H. Arnold, was taught to fly in May 1911 at the Wright Company's flying school at Huffman Prairie, now part of Wright-Patterson Air Force Base.

America's first military aircraft engine, the Curtiss four-cylinder water-cooled engine, was used in the 1908 Signal Corps Dirigible No. 1. Developing about twenty-five horsepower, it drove a tubular steel shaft twenty-two feet long on which was mounted a wooden propeller designed by Lt. Thomas E. Selfridge. In the official speed trial the Baldwin airship reached 19.61 miles per hour.

Tragedy strikes in the form of death for Lt. Thomas E. Selfridge, who gained fame of sorts on September 17, 1908, as the first man killed in an aircraft accident. The pilot, Orville Wright, barely escaped with his life. The propeller shown in the case (*bottom left*) was on that 1908 Flyer. Pictures of the 1909 Wright Flyer are of the trial flights prior to the plane's being accepted and designated as Signal Corps Airplane No. 1. A diorama depicting an early flight is in the window to the right.

The museum has several dioramas such as the one depicted here. (A diorama is a three-dimensional picture with small figures placed so they blend with the painted background.) The skeleton tower is part of the system to catapult the Flyer down a track and into the air. The aircraft was moved on the ground by means of the wheels shown near the tower. This scene depicts the Wright 1909 Flyer at Fort Myer, Va.

Visitors are encouraged to touch this exhibit and learn the four Principles of Flight: drag, gravity, lift, and thrust. The video display also presents how to control an airplane in flight.

Lt. Benjamin D. Foulois transported the 1909 Flyer to Fort Sam Houston and taught himself to fly it even though he had never made a solo flight, takeoff, or landing. On March 2, 1910, he made his first flight and by September had flown the plane on sixty-one "hops." In 1911 the plane was retired from service.

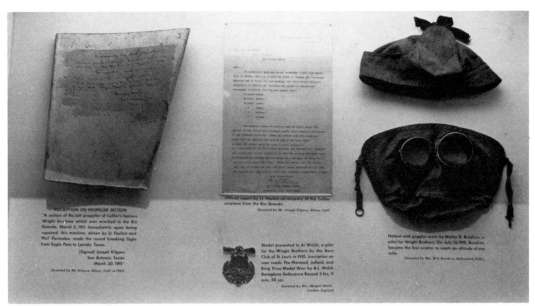

Following the retirement of the 1909 Flyer, Mr. M. R. Collier, owner of *Collier's Magazine,* loaned the Army a new 1910 Wright "B" airplane. Pictured are a section of a propeller (*left*) from Collier's plane, which was flown by Lt. Foulois, and helmet and goggles worn by Walter R. Brookins, a pilot for the Wright brothers. In 1910 Brookins became the first aviator to reach an altitude of one mile.

Two youngsters get a closeup view of a model airplane. Displays around them feature the first weapons fired from aircraft and pilots training at early flying schools.

The first shot fired from an airplane occurred on August 20, 1910, at Sheepshead Bay racetrack near New York City. With Glenn Curtiss piloting, Lt. Jacob E. Fickel fired a rifle at a small target from an altitude of 100 feet.

Lt. Myron Crissy and Mr. Phillip O. Parmalee demonstrating the first drop of a live bomb, January 15, 1911, near San Francisco. The encased tennis ball object is one of several imitation bombs made for the Harvard-Boston Aero Meet in September 1910. Claude Grahame-White demonstrated the possibility of destroying ships by dropping bombs down their funnels.

The second airplane purchased by the Army Signal Corps was the Curtiss 1911 Model D Type IV pusher, similar to the Curtiss Standard D pusher. It was one of five airplanes ordered by the Army that year and had a top speed of 50 mph. The wings were made in sections so that the craft could be disassembled and transported by wagon. This Model D reproduction was carefully made in 1985-87 by highly skilled museum personnel who relied heavily on early photographs and an existing factor-built Curtiss pusher. Except for the engine, all materials are essentially the same as used in the original.

The Army's first permanent aviation school developed from the Glenn Curtiss flying school on North Island in San Diego Bay. Three officers stationed in California were ordered to San Diego in 1911 as the first students. Later, Signal Corps Airplane No. 50, seen here, was equipped for airborne radio experiments at North Island.

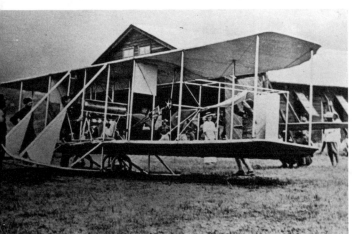

With Lt. Frank P. Lahm of the 7th Cavalry Regiment in charge, the United States opened a flying school in the Philippines on March 12, 1912, for Americans. Airplane No. 13 of the Wright "C" Series, seen here, was destroyed on its first flight with pontoons.

Further seaplane flights were conducted in the Philippines after this Burgess hydroplane, beginning takeoff run, was delivered in September 1913, but they ended in January 1915 when Lt. Herbert A. Dargue wrecked it.

Wall exhibits of memorabilia and photographs line the chronological walk through the museum. Items from the 1st Aero Squadron, created in early 1913, are seen here on the right. The enlargement from the adjoining panel shows Lt. Carl Spaatz wearing a football helmet while learning to fly in 1916 at San Diego's North Island; the airplanes here are lined up for Saturday inspection.

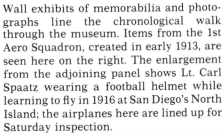

Revolutionary Leader Pancho Villa crossed the border from Mexico and raided Columbus, New Mexico, on March 9, 1916, killing seventeen Americans and destroying part of the town. Capt. Benjamin D. Foulois arrived there March 15 with the 1st Aero Squadron to support Brig. Gen. John J. Pershing's 15,000 ground troops. High winds, dust storms, and snowstorms thwarted the flying efforts.

Army troops survey the damage caused by Pancho Villa's raid.

The 1st Aero Squadron used these Curtiss JN3 airplanes in Mexico. Originally the squadron there consisted of eight planes. Within a few weeks only two were operational, and these were soon condemned.

In 1916 and 1917 Curtiss produced R-3 and R-4 airplanes for the Signal Corps, some of which were used by the 1st Aero Squadron during the Pershing Punitive Expedition into Mexico. These R-3 and R-4 airplanes were powered by this type Curtiss V2-3 engine. It was replaced during World War I by the Liberty, a much more efficient engine.

The Wright 6-60 liquid-cooled engine was used by the Army in 1912/13 to power its Wright Model "C" and "D" airplanes. The 6-60 had six cylinders and was designed to produce sixty horsepower.

Museum personnel also restored this Standard J-1, which resembles the Curtiss JN-4D Jenny. The basic version of the J-1 was used as a trainer from 1916 to 1918. Standard was joined by the Dayton-Wright, Fisher Body, and Wright-Martin companies in building 1,601 of these two-seat primary trainers.

These LWF trainers were lined up for inspection at Mineola, N.Y., in the spring of 1917. On April 5, 1917, there were fifty-six airplane pilots and fifty-one flying students in the Aviation Section. It had fewer than 250 planes, all trainers and not one suitable for carrying machine guns or bombs on a combat mission. Although war with Germany was but one day away, the Aviation Section was a least two years behind Germany in practically all aspects of military aviation. An enormous job lay ahead. The flying school in San Diego had recently been supplemented by additional schools at Mineola, N.Y.; Chicago; Memphis; and Essington, Pa. The Memphis flying school was located inside a race track. Large tents served as hangars.

★ *World War I*

World War I brought a great deal of destruction to Europe from July 1914 to November 1918. Because of its isolationist policy, the United States was not directly involved until June 1917, when the first of the American Expeditionary Force landed in France. Though the Air Force Museum World War I section deals with some of the destruction, it emphasizes the men and planes of the aerial war. Artifacts, equipment, personal diaries, and photographs of legendary pilots are displayed on wall panels. Authenticity is a must, and it has been attested to by veterans of that war who have visited the museum. Of course, the eye-catching airplanes of that era are also featured. When compared later in the museum with the sleek jet aircraft, the fragile planes of WW I do seem to be mere toys.

The United States declared war on Germany in April 1917. At that time the Signal Corps' Aviation Section had 131 officers, practically all pilots and student pilots, 1,087 enlisted men, and less than 250 airplanes. And none of these planes was worthy of joining the air battle. Even with the appropriation of $640 million for aeronautics by a then enthusiastic Congress, American industry was unable to produce one American-designed airplane that saw combat. With only a few airplane plants and hardly more than a dozen aeronautical engineers, the nation's resources in April 1917 were not a broad enough base on which to build and equip an effective air arm rapidly. Using a British design, United States industry did produce 3,431 DeHavilland DH-4s prior to the Armistice. Most were built by the Dayton-Wright Company, and 417 got into combat.

The first aerial force sent to France began arriving in Europe in September 1917. The first planes used by the First Aero Squadron on combat flights were

AR-1s. It was April 1918 before they shot down their first sky opponent. Much of the intervening time had to be spent in training. During the final seven months of the war, the gallant men who were to inspire generations to come shot down 756 enemy airplanes and 76 balloons, while losing 289 airplanes and 48 balloons to enemy fire. And they did much of the job with second-rate equipment.

Out of the combat came a new warrior—the ace—victor of five or more air battles. The United States had thirty-one, topped by Capt. "Eddie" Rickenbacker, the American Ace of Aces and recipient of the Medal of Honor. Career highlights of many of these heroes are depicted as the museum visitors continue their chronological walk through aviation history in the Early Years Gallery.

For nearly two years before the United States entered World War I, a number of American pilots voluntarily fought the Germans as part of the French Aviation Section. Originally known as the *Escadrille Americaine*, they changed their designation to the Lafayette Escadrille when the Germans complained. By February 1918, they had downed fifty-seven enemy aircraft while suffering nine of their own pilots killed. Raoul Lufbery and Bill Thaw perhaps were the best known of the Lafayette Escadrille.

The *Croix de Guerre* and the Legion of Honor awarded to H. S. Jones by France are displayed here. Jones is pictured with the Escadrille Lafayette standing third from the right in the accompanying photograph taken in July 1917. The legendary Lufbery is seated fourth from the right holding Whiskey (a lion), one of their mascots.

To Air Force Museum and best wishes to all

These French decorations were won by Eugene Bullard, the first black American to serve as a military pilot. He flew with the French Aviation Section and remained in France until World War II.

The famous "Red Knight of Germany," more commonly known as the "Red Baron" (Manfred von Richthofen), scored most of his eighty victories while flying the Albatros fighter. He is more closely associated with the Fokker Dr. 1 triplane, which he flew during the latter part of his career.

The Fokker Dr. 1 triplane, which the Red Baron flew on various occasions during World War I, was exhibited in a museum in Berlin in the 1930s. Reportedly, Hitler refused to give permission for it to be moved to a place of safety during World War II, and it was completely destroyed by Allied air attacks on the German capital.

A variety of airplanes were used in combat long before the United States entered the war in 1917. This is a Voisin Type 5 airplane with Hotchkiss gun as photographed in 1915.

Rumpler C-IV reconnaissance airplane with flexible Parabellum gun in 1917.

British two-seater pusher airplane with swivel machine gun in front for use by the observer. This plane was shot down and captured; the victorious German pilot stands in the rear cockpit.

When the United States entered World War I, it had no military air arm capable of fighting an enemy. Although Congress appropriated $640 million for aeronautics, the United States eventually had to purchase most of its combat airplanes. Despite early failures, United States aircraft production became an outstanding success with the nation producing 50 percent more airplanes in the nineteen months it was in the war than Great Britain had produced during its thirty-one months of war.

America's greatest technological contribution was the development and mass production of the twelve-cylinder Liberty engine. It was the mainstay of the Air Service for ten years after the war. Seen here is Maj. H. H. Arnold with the first Liberty 12 engine.

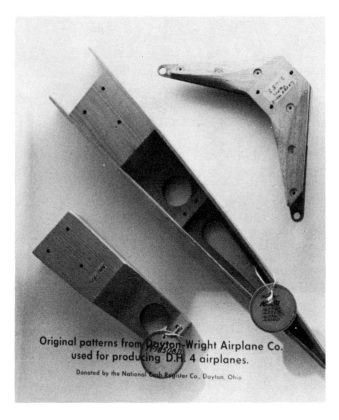

Original patterns from Dayton-Wright Airplane Co. used for producing D.H. 4 airplanes.
Donated by the National Cash Register Co., Dayton, Ohio

King George V welcomed United States troops to the British Isles with a handwritten letter that accompanies this scene in a display panel. The 211th Aero Squadron is shown arriving there on August 1, 1918.

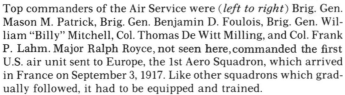

Top commanders of the Air Service were (*left to right*) Brig. Gen. Mason M. Patrick, Brig. Gen. Benjamin D. Foulois, Brig. Gen. William "Billy" Mitchell, Col. Thomas De Witt Milling, and Col. Frank P. Lahm. Major Ralph Royce, not seen here, commanded the first U.S. air unit sent to Europe, the 1st Aero Squadron, which arrived in France on September 3, 1917. Like other squadrons which gradually followed, it had to be equipped and trained.

Major Fiorello H. La Guardia, standing between two Italian officers, commanded United States Air Service personnel in Italy. He later gained fame as mayor of New York City from 1933 to 1945. The sixty-five Americans who flew with Italian aircrews in Italian airplanes represented about 25 percent of the Italian combat pilot force. The Italian primary flying course at Foggia graduated 406 U.S. cadets, most of whom were immediately transferred to France.

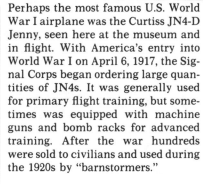

Perhaps the most famous U.S. World War I airplane was the Curtiss JN4-D Jenny, seen here at the museum and in flight. With America's entry into World War I on April 6, 1917, the Signal Corps began ordering large quantities of JN4s. It was generally used for primary flight training, but sometimes was equipped with machine guns and bomb racks for advanced training. After the war hundreds were sold to civilians and used during the 1920s by "barnstormers."

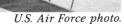

U.S. Air Force photo.

During August 1918, German pilots flying the superior Fokker D. VII shot down 565 Allied aircraft. It first went into combat in May 1918. The reproduction shown here was placed on display in May 1996. In the background is a Consolidated PT-1 trainer, the first American airplane purchased in quantity after the war.

The Spad VII, a French-designed fighter airplane, made its initial flight in July 1916. It showed such promise that it was put into production at once and used by both French and British combat squadrons. The Lafayette Escadrille was equipped with these in February 1918 when it joined the United States Air Service. This plane was restored and flown in 1962-66 at Selfridge Air Force Base, Michigan.

Capt. "Eddie" Rickenbacker, the U.S. Ace of Aces, and a few other American pilots became well known to the American public during what has been termed the "golden age" of individual aerial combat. This famous pose was made of Rickenbacker and his Spad XIII at an aerodrome near Rembercourt, France, in September 1918. *U.S. Air Force photo.*

Captain Rickenbacker was credited with downing twenty-two enemy planes and four balloons, but probably shot down more. He was later presented the Medal of Honor by President Hoover. *U.S. Air Force photo.*

The famous Hat-in-the-Ring insignia from Captain Rickenbacker's Spad 13 is displayed at the museum along with others cut from aircraft by World War I crews. Souvenir snapshots surround the insignia. Rickenbacker's autograph appears in the base of the figure one.

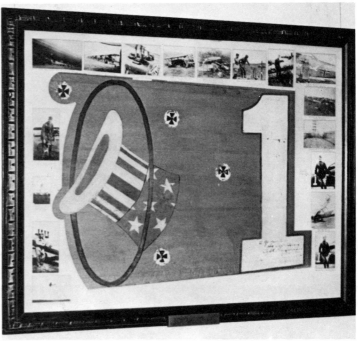

When the new Air Force Museum was dedicated in 1971, Captain Rickenbacker was among the honored guests. He posed with his portrait for the *Fairborn Daily Herald.*

Dubbed "Tommy" by the pilots who flew it, the Thomas-Morse S4C Scout was the favorite single-seat training plane produced in the United States during World War I. It was used by almost every pursuit flying school in the country during 1918. Some were still being used in the mid-1930s for World War I movies filmed in Hollywood. Today, several are still being flown by aviation buffs.

Brig. Gen. William "Billy" Mitchell flew this SPAD XVI as his observation and command vehicle during the latter days of the war. At the time, the SPAD XVI was one of the fastest and most maneuverable two-seaters in combat. This aircraft is on loan from the National Air and Space Museum. *U.S. Air Force photo.*

One of the world's earliest strategic bombers, the Caproni Ca. 36 was painstakingly restored in 1988-90 at the Air Force Museum. During World War I, it could carry an internal bomb load of 1,800 pounds at 70 to 80 mph. Some Americans who learned to fly in Italy flew this type of aircraft, thus gaining America's early experience in strategic bombardment.

World War I aerial combat and machine guns used in aerial duels are featured in this display. The three-dimensional scenes across the top (*left to right*) are in close-up: D.H. 4s under attack by German Fokker D. VIIs, German observation balloon being attacked by an S.E. 5, and a Spad 13 victorious over a German L.V.G. observation plane. Suspended to the left of the main display is a scaled model of the Salmson 2A.2, an extinct observation plane. The model was commissioned in 1968 by the late Eugene W. Kettering and built over an eighteen-month period by Joseph Fallo of Dayton. It features control surfaces that can be worked from the cockpit.

This actual Salmson 2A.2 was the model for the one on display. The United States procured 705 of these from the French for use by ten Air Service observation squadrons. *U.S. Air Force photo.*

During World War I Germany relied primarily on two types of machine guns to arm its airplanes. One was the Parabellum 7.92 mm (*right*) that young Anthony Fokker used in 1915 to develop the first successful system for "synchronizing" a machine gun to fire through a revolving propeller. In the 1916–1918 period, the Spandau 7.92 mm (*left*) was used almost exclusively as a fixed fuselage gun for firing through the propeller while the Parabellum was used primarily by observers as a "flexible" gun on a swivel mount.

Hatched in the trenches in France during 1918, "Stumpy John Silver" was a veteran message carrier by the time he was a few months old. When he was eleven months old, he flew into a furious German bombardment. Miraculously, he arrived twenty-five minutes later, having averaged a mile a minute, at his destination. His body was torn, but the message tube hung intact by the ligaments of his missing right leg. He was returned to health but not to battle. He was assigned to the 11th Signal Company and lived to be nearly eighteen years old at Schofield Barracks in Honolulu. His name is called on each Organization Day of the 11th Signal Company, and the senior noncommissioned officer answers, "Died of wounds received in battle in the service of his country." He stands as a symbol of the heroic and faithful service of pigeons to our combat forces during World War I.

This map indicates a wide area of combat in 1918. To the right two American observers report enemy movements to their ground commander while suspended in a basket from a Caquot Type R balloon (*see pages 68-69*). Below them is the nose of a British Sopwith Camel (*page 64*), the most successful fighter of the aerial campaign. Over the map hovers a Halberstadt CL IV (*page 66*), a main ground attack airplane for the German Army.

Regulations required Air Service officers to wear "choke" collars. But using the excuse that the prescribed collars made their necks sore from constantly turning their heads while flying, American pilots wore the British-style open-collar blouse and trousers.

Few Sopwith Camels remain today from World War I. When this F.1 exact replica was "rolled out" in December 1974 at the Air Force Museum, the second-ranking American ace to survive the war was present. Ace George Vaughn flew Camels with the 17th Aero Squadron and was credited with thirteen combat victories. He listened to the restored engine and smelled the castor oil fumes. "It's the real thing," he said.

Museum employees built their British Camel using original factory blueprints marked "confidential" and "secret." A number of the parts are from World War I. These include the 130-horsepower French Clerget rotary engine, two Vickers machine guns, George Vaughn's gunsight, and the tires and wheels.

On the fourth pull of the propeller, the Sopwith Camel's engine fired into operation. Walter J. Olsen, who headed the construction project, had the honor of sitting in the cockpit. As a safety precaution, the aircraft tail was tied down.

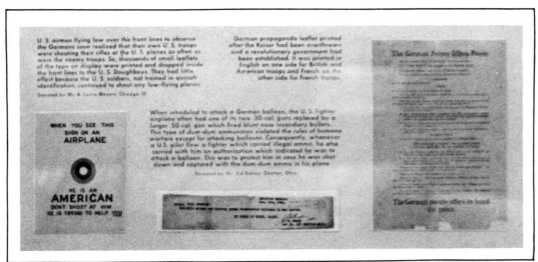

Identification posters of German airplanes hung in ready rooms for U.S. pilots to study so they could distinguish enemy from friendly aircraft. But there was another problem. American troops on the ground fired at all low-flying planes. To remedy that problem, U.S. airmen had leaflets *(left)* printed for the doughboys. They read: "When you see this sign on an airplane, he is an American. Don't shoot at him. He is trying to help YOU." The leaflets had little effect because the ground troops were not trained in aircraft identification. A German propaganda leaflet is to the right.

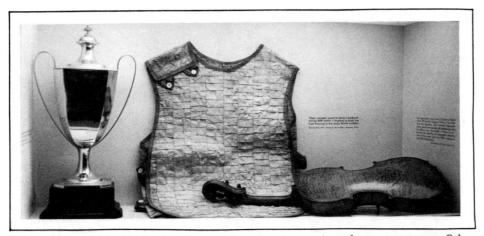

Winters are cold in England and some Americans were issued paper sweaters. Other mementos brought home include the loving cup which had been presented to the commander of the 21st Aero Squadron and a French violin which had been autographed by Capt. Eddie Rickenbacker. The violin provided entertainment at U.S. flying fields.

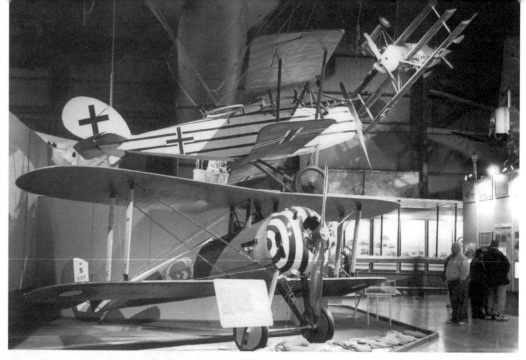

Wood and hardware from an original Nieuport 28 (N.28C-2) were used by museum personnel in the reproduction shown on the display floor. Many American aces flew such French-built fighters in World War I. On the wall over it is a German Halbertstadt CL IV, restored for the museum in Berlin by the Museum fur Verkehr und Technik. In early 1918, the Halberstadt was very successful in attacking Allied positions ahead of advancing German troops. Diving from above is a reproduction of the Fokker Dr. 1, made famous by Germany's Rittmeister Manfred von Richthofen—the "Red Baron." This nimble Fokker became known as the best "dogfighter" of the war.

Museum visitors sometimes see themselves pictured in a display and are invited by the staff to autograph the photographs. Such was the case when Stephen W. Thompson of Dayton visited the museum on October 15, 1973. That's his uniform and picture in the top left. While visiting a French bombing squadron on February 5, 1918, he became the first American to shoot down an enemy airplane in World War I. Capt. William C. Lambert, America's No. 2 Ace of the war, is featured in the bottom half. While flying with the Royal Air Force in World War I he was credited with 21½ "kills," 4½ less than Capt. "Eddie" Rickenbacker. Lambert served with the USAF in World War II.

Lt. Quentin Roosevelt, the youngest son of former President Theodore Roosevelt, died when his Nieuport 28 was shot down behind German lines on July 14, 1918. The Germans placed a crude cross (*right*) over his grave. On July 12, 1944, his brother, Brig. Gen. Theodore Roosevelt, Jr., died of a heart attack in Normandy following the World War II invasion of France. The two brothers are now buried side by side at Omaha Beach in France.

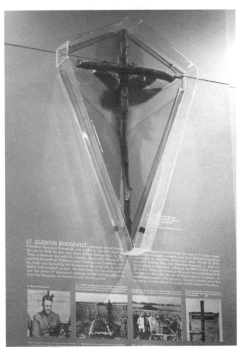

Heavy clothing, boots, goggles, and a padded helmet were worn by pilots to keep warm in the air. The display card in the lower left contains pieces of fabric from Lt. Quentin Roosevelt's airplane. Next to it is an original German postcard printed and sold showing the body of Lieutenant Roosevelt beside his crashed Nieuport.

Lt. David E. Putnam (*photograph*) was the leading ace of the Air Service when he was shot down and killed on September 13, 1918. He had downed twelve aircraft. A control stick from a downed Fokker is displayed over a mask which was worn by Allied flyers to protect them from cold air blasts at high altitude. First aid kits are shown bottom right.

Members of the U.S. Balloon Section made 1,642 ascensions and were aloft for 3,111 hours observing enemy activities. Although thirty-five balloons were shot down by German planes and another twelve destroyed by enemy ground fire, only one American observer was killed in 116 parachute jumps that were made. He died when pieces of his burning balloon fell on his descending parachute. The rectangular item on the left wall is a part of that balloon. Across the top is an observer's logbook for April 14 through November 15, 1918.

A long-range aerial camera being prepared for an ascent.

This is what the French landscape looked like behind the front lines.

Caquot Type R balloons such as this one were used during World War I as observation platforms by the Allies. It is ninety-two feet long and thirty-two feet wide. Members of the National Association of American Balloon Corps Veterans discovered it in possession of the Royal Aircraft Establishment in England, which donated it to the museum in 1976. The Goodyear Aerospace Corporation, which had manufactured such balloons in World War I, assisted museum personnel in the restoration. *U.S. Air Force photo.*

The world's first "guided missile," the Kettering Aerial Torpedo, was invented by Charles F. Kettering and built in Dayton in 1918 for the Signal Corps. The Kettering "Bug" took off from a dolly running along a track and was guided toward its target by a system of internal preset controls. At the predetermined time the engine shut off, the wings were released, and the "Bug" plunged to earth where its 180 pounds of explosives detonated upon impact. Less than fifty "Bugs" were completed at the time of the Armistice. The "Bug" on exhibit is a full-size reproduction built by museum personnel. Its wings stretch nearly fifteen feet.

One hundred and twenty-three flyers of the Air Service were captured and placed in German prisoner-of-war camps. Other American POWs included two balloonists, nineteen Americans flying with the British, ten with the French, one with the Italians, and one enlisted man who drove too close to the front. The POW camp at Villingen is at top left. Russian POWs decorated the movie theatre there (*center*). A funeral procession and burial of an American POW are shown across the bottom.

At the end of World War I an independent Poland was created from territory previously held by Germany, Austria, and Russia. Poland thus regained the autonomy she had lost in 1831. Almost immediately the new Polish republic was invaded by the Bolsheviks. Two World War I combat pilots, Merian C. Cooper and Cedric E. Fauntleroy, received permission to recruit former U.S. flyers for a Polish Squadron similar to the Lafayette Escadrille. Seventeen Americans volunteered and served in the Kosciuszko Squadron, named in honor of Tadeusz Kosciuszko, the Pole who had fought in the American Revolution under George Washington. The Bolshevik invasion ended in May 1921 with victory for the Poles. The occupation of Germany is shown below.

★ 4 | *Era Between WWI and WWII*

Out of World War I came hard-earned experience, new tactics and equipment, and a place for the airplane in the military structure. But the years between the World Wars became frustrating ones for the enthusiasts of military aviation. The more thoughtful air leaders had come out of the war with the belief that airpower would be the dominant weapon of the future. They had to prove their theories while handicapped by a shortage of funds. Thus they turned to staged events and spectacular feats to prove their theories.

Brig. Gen. "Billy" Mitchell demonstrated the vulnerability of seapower to air attack when in 1921 his airplanes sank three captured German ships and an obsolete and stripped American battleship. In another demonstration in 1923 his Air Service bombers sank two more obsolete U. S. battleships. To some, the lesson wasn't fully understood until the Japanese devastated the United States Navy at Pearl Harbor on December 7, 1941. Other demonstrations included coast-to-coast flights, speed dashes in flimsy pursuit aircraft, altitude records in open-cockpit biplanes, aerial refueling, and flight endurance records. Because of America's vast distances, endurance and long-range flights were emphasized.

The first in-flight refueling occurred in 1923. The next year a Curtiss pursuit plane flew across the country in twenty-two hours. Also in 1924 two Douglas World Cruisers completed the first aerial circumnavigation of the earth. Flying at a top speed of 100 mph and covering 26,345 miles, the trip took 175 days to complete. But only fifteen days of actual flying time were logged. A wall exhibit built around the accomplishments of those World Cruisers dominates that portion of the Air Force Museum dealing with the period between the World Wars. In a less spectacular setting, immediately after WW I, military pilots flew airmail routes with rebuilt D.H. 4s from the war. Many civilian pilots purchased surplus Jennys and they soon became the mainstay of colorful barnstorming tours and aerial circuses.

It was not until the 1930s that the legacy of WW I equipment was lifted from the Air Corps. Steady improvements in aircraft design and performance revived the hopes of airpower advocates. Congress in 1926 had recognized the concept of military aviation as an "offensive striking force" rather than a mere auxiliary service. The development of bombers was so accelerated, however, that pursuit aviation fell far behind. Limited funds did not permit the development of both fighters and bombers. It took the uneasy rumblings of another war in Europe to get the skeptics to look to the sky.

Visitors to the museum can relish the experiences of the open-cockpit pilots of the 1920s and 1930s as they examine and marvel at authentic airplanes from this era as they continue through the Early Years Gallery.

Affectionately called the "Peashooter" by its pilots, the Boeing P-26A was the first all-metal, monoplane fighter produced for the Air Corps. And it was the last Air Corps aircraft accepted with an open cockpit, fixed undercarriage, and externally braced wing. First flown in 1932, the P-26 was the corps' frontline fighter until replaced in 1938-40 by the Curtiss P-36A and Seversky P-35. In the left background, a mechanic works on a Boeing P-12E *(see page 81)* while a Stearman PT-13D Kaydet looms overhead. The PT-13 was typical of the biplane primary trainer during the late 1930s and in WW II. This Kaydet was donated to the museum in 1959 by the Boeing Company, which in 1938 had purchased the Stearman Company.

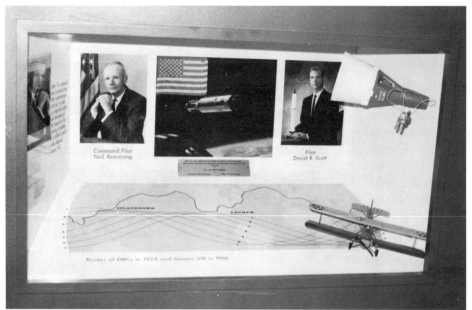

Gemini 8 astronauts carried pieces of a World Cruiser aboard their capsule when they circled the earth in 1966.

Pontoons were fitted on the World Cruisers near Seattle for their flight across the Pacific and along the Asian coast. They were not replaced by wheels until the planes had reached Calcutta, India.

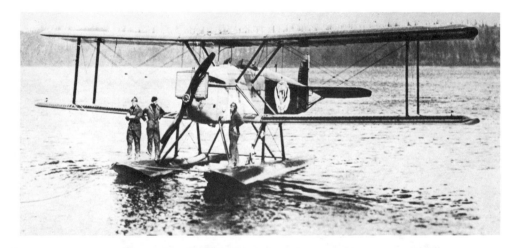

For their tremendous achievement in flying around the world, the World Cruiser flyers were awarded the coveted Mackay Trophy "for the most meritorious flight of the year." Memorialized in bronze are (*left to right*) Lieutenants John Harding, Jr., Erik Nelson, Leslie Arnold, Lowell Smith, Henry Ogden, and Leigh Wade.

"ROUND-THE-RIM FLIGHT"

In its desire to test the long-range capabilities of the airplane, the Air Service decided to fly a Glenn Martin bomber (GMB) completely around the periphery of the U.S. The flight, which began at Bolling Field, Washington, D. C. on July 24, 1919, was made in a counterclockwise direction. Since time and speed were not factors, the flight proceeded leisurely westward across the northern states, down the Pacific Coast, and eastward along the Mexican border and across the southern states, arriving back at Bolling on Nov. 9, 1919. The total distance of approximately 10,000 miles was flown in 114 hours, 45 minutes. This was a tremendous achievement for such an early period in the development of the multi-engine bomber.

Before the round-the-world flight in 1924, the Air Service in 1919 had a crew of four fly a Glenn Martin bomber around the periphery of the U.S. to test long-range capabilities. From July 24 to November 9, they leisurely flew 10,000 miles in 114 hours, 45 minutes. Pictured above (*left*) are Lt. Col. R.S. Hartz, Sgt. Jerry Dobias, Sgt. Jack Harding and Lt. Ernest E. Harmon.

These nine and ten-foot propellers, as well as a great deal of other equipment, were designed and/or tested at Dayton's McCook Field in the 1919-22 era.

Because McCook Field was too small, in 1923 the "Barling Bomber" was tested at Wilbur Wright Field, now Patterson Field. Everything about the XNBL-1 was large, including its nose wheel and main landing gear wheel.

Fifty versions of the English S.A. 5a, then known as the SE-5E, were assembled during 1922-23 in the United States by the Eberhart Steel Products Company from spare parts it had purchased. It was acquired later for the museum through a donation from the estate of Lt. Col. William C. Lambert and the Air Force Museum Foundation. Colonel Lambert, a World War I ace with 21½ victories, flew such fighters as an American member of the Royal Air Force.

The Glenn Martin bomber, built in October 1918, was too late for World War I. But it was the Air Service's first truly successful multiengine bomber design. This improved 1920 version, the MB-2, set the basic design for U.S. bombers for the decade. Using these improved bombers, Brig. Gen. Billy Mitchell arranged demonstrations in 1921 and 1923 in which obsolete German and American ships were sunk. *U.S. Air Force photo.*

In the years after World War I, more than eleven hundred obsolete D.H. 4Bs saw wide use with the Air Service, including carrying airmail for the Post Office Department. In June 1923 the Air Service successfully demonstrated in-flight refueling and two months later set a world endurance record by having a D.H. 4 stay aloft over San Diego for thirty-seven hours and fifteen minutes. A reproduction of the type D.H. 4B used in 1919-21 to patrol the Mexican border is displayed at the museum. *U.S. Air Force photo.*

This aviation routing beacon was the property of the U.S. Air Mail Service of the Post Office Department. Its purpose was succinctly stated on its base: "Pilots depend on this beacon for their safety."

The Verville-Sperry M-1 Messenger was designed in 1919/20 at McCook Field, Dayton. It was intended to serve as an aerial dispatch carrier and to maintain liaison between field units. The hook mounted above the upper wing was used in the first successful mating of an airplane and an airship while in flight, on September 18, 1923, over Langley Field, Virginia. Lt. Rex K. Stoner flew his Messenger under a D-3 airship and "landed" on a waiting trapeze.

This Douglas 0-38F observation plane was one of the first military aircraft assigned to Alaska, landing at Ladd Field near Fairbanks in October 1940. The following June, because of engine failure, it crash landed in the wilderness. Although the pilot, Lt. Milton H. Ashkins, and the mechanic, Sgt. R. A. Roberts, hiked to safety, the 0-38 remained in the woods until June 1968 when it was brought out by helicopter. Its rescue and restoration are told by the movie box at the right.

With a welded fuselage framework of steel tubing, the Consolidated PT-1 was so sturdy and dependable that it was nicknamed the "Trusty." Procured in 1925, it was the first airplane purchased in quantity (221) since World War I. The Trusty was used extensively into the early 1930s to train pilots in California and Texas. This one was obtained in 1957 from Ohio State University.

This unique observation amphibian with an inverted Liberty engine is the Loening OA-1A, seen here in two views. Used extensively in the Hawaiian and Philippine Islands, the plane combined features of both a landplane and seaplane by merging the fuselage and hull into a single structure. Capt. Ira C. Eaker and Lt. Muir S. Fairchild piloted the *San Francisco* on a 22,000-mile Pan American good will tour of twenty-five Central and South American countries from December 21, 1926, to May 2, 1927. Four other OA-1As made the tour for which all the fliers received the Mackay Trophy and Distinguished Flying Cross.

Development of the Liberty engine (*bottom right*) has been cited as the nation's outstanding contribution to aeronautics during World War I. But so many were built that the U.S. was saddled with them for more than a decade, hindering further development and procurement of new engines. Congress responded by prohibiting the use of the Liberty engine in new aircraft, thus aiding industry and the Air Corps.

More answers to aerial refueling questions were supplied in 1929 when Maj. Carl Spaatz, Capt. Ira C. Eaker, and Lt. Elwood R. Quesada flew their Fokker C-2 for an endurance record of nearly 151 hours. They flew over the Los Angeles area between January 1 and 7 in their "Question Mark," making forty-three hookups and perfecting the refueling techniques developed in 1923. In later years, Spaatz, Eaker, and Quesada were to contribute much to the development of airpower and the Air Force. *U.S. Air Force photo.*

Although never used in combat, the Curtiss P-6E Hawk is remembered as one of the most beautiful biplanes ever built, seen here at the museum and in flight. It was a first-line pursuit aircraft of the early 1930s and was the last of the fighter biplanes built in quantity for the Army Air Corps. Despite its excellent performance, only forty-six P-6Es were ordered because of the shortage of funds during the austere days of the Depression. The museum's P-6E may be the only one of its kind in existence. It was restored by Purdue University.

One of the best-known Air Corps fighters between the World Wars, the Boeing P-12 was first flown by military pilots as the Navy XF4B-1. This P-12E, restored at the museum, was powered by a 525-horsepower radial engine. P-12s were flown by pursuit squadrons from 1929 until replaced by Boeing P-26s in 1935. A Stearman PT-13D is shown overhead.

Distinguished visitors are not uncommon at the museum. Here Lt. Gen. (Retired) James H. Doolittle *(center)* relates a personal observation to aviation history. Newscaster Lowell Thomas, historian for the Douglas World Cruisers' flight, is on the right.

This is the instrument panel that Lt. James H. Doolittle used on September 24, 1929, when he made history's first blind flight. Years later, on April 18, 1942, Doolittle led the first U.S. bombing raid on Japan.

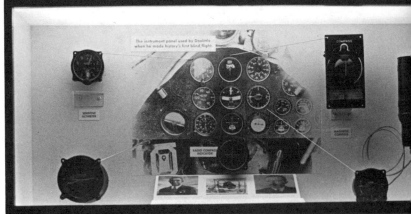

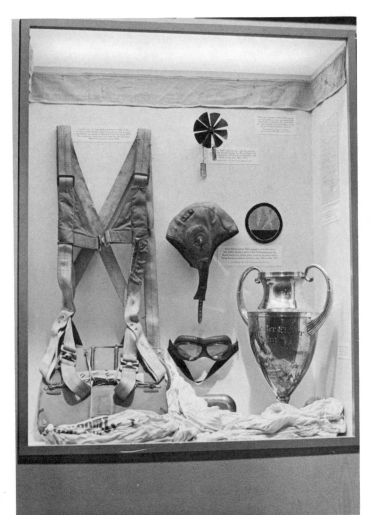

An improved Switlik type S-1 seat pack parachute *(left)* was used in the 1930s. It was a considerable improvement over the original parachutes designed at Dayton's McCook Field after World War I. The Army flying helmet and Navy goggles were worn aboard the Keystone biplane bomber in 1932 and 1933.

Major Thurman H. Bane piloted the first flyable helicopter on December 18, 1922. He kept it aloft for nearly two minutes at a height of six feet. A model of the multi-bladed machine is displayed in the case. The helicopter was designed at McCook Field by Dr. George De Bothezat and Ivan (Eremeeff) Jerome.

Medals, uniforms, and personal effects of Brig. Gen. "Billy" Mitchell, the outspoken advocate for air power, are displayed along with aerial photographs of his accomplishments. The Mark I 4,300-pound bomb (*right*) was developed for the Mitchell bombing trials on obsolete ships and tested in September 1921. However, the largest bomb he was permitted to use was a 2,000-pounder.

Mitchell's bombers sank three captured German vessels and three obsolete American battleships in 1921 and 1923 to prove that airplanes could destroy naval ships. *U.S. Air Force photo.*

This diorama depicts the LePere LUSAC-11 high altitude research airplane over McCook Field. Many aviation achievements were recorded there in the early 1920s before the Dayton facility was closed and moved approximately ten miles to the new Wright Field. In 1989 the U.S. Air Force Museum traded a P-38 to a French museum for the last remaining LUSAC-11.

THE LePERE LUSAC-11 HIGH-ALTITUDE RESEARCH AIRPLANE OVER McCOOK FIELD

Capt. Georges LePere of the French Air Force came to America to design the Packard LePere LUSAC-11, which was built in 1918 in Detroit, for the U.S. Air Service. Three were shipped to France for combat evaluation as the war ended. Later, others were used over Ohio for high-altitude research.

Ten Martin B-10 bombers commanded by Lt. Col. Henry H. Arnold in 1934 flew from Washington, D.C., to Fairbanks, Alaska, and return. The B-10 was heralded as the air-power wonder of its day. It was the first all-metal monoplane bomber to be produced in quantity and could cruise at 183 miles per hour with 2,200 pounds of bombs carried internally. The aircraft on display was sold to Argentina in 1938 and donated by that country's government to the United States for the museum in 1971. This last-known remaining B-10 was painstakingly restored from 1973 to 1976 at Kelly Air Force Base, Texas, by members of the Air Force Reserve and other base volunteers.

This model of the Keystone B-4A is representative of those flown out of March Field, California, in 1932 and in the Philippines in 1936. These and other Keystone light bombers were replaced by the B-10, which was replaced by the Boeing B-17 and Douglas B-18. George Lee made this model from scratch over a twelve-year period. It was donated to museum in 1997 by Mrs. Milly Lee of San Leandro, California.

During Air Corps bomber trials at Wright Field in 1935, the less-costly Douglas B-18 prototype was chosen over the forerunner of the Boeing B-17, which many believed was superior. This was in keeping with a policy of getting the maximum number of new weapons with the funds available. By early 1942, the B-18 Bolo was being used as a transport and for anti-submarine duty. The B-18A on display was stationed in 1939-42 at Wright Field.

The only known surviving Seversky P-35 was restored by the Minnesota Air National Guard and presented to the Air Force Museum in mid-1974. This forerunner of the Republic P-47 was the first single-seat, all-metal pursuit plane with a retractable landing gear and enclosed cockpit to go into regular service with the Air Corps. Seventy-six were received in 1937/38 and sixty more designated as P-35As, in 1940. Twenty two-seat versions were sold to the Japanese Navy in 1938. These became the only American-built planes used operationally by the Japanese in World War II.

★ 5 World War II

Airpower was a proven factor, proven by Hitler's Luftwaffe dive bombers, long before the United States was pushed into World War II. If there were any doubts, the Japanese erased them on December 7, 1941, with their sneak air attack on Pearl Harbor in Hawaii. But military airmen and leaders of the aviation industry had planned for the eventuality of war, and America met the challenge. Many of the aircraft which made victory possible for the Allies are on display at the Air Force Museum, as well as a number of German, Italian, and Japanese aircraft. The gallant men who flew and maintained American and Allied aircraft are also remembered, particularly in new life-size dioramas, with history being related in terms of human involvement.

Personal items from the B-25 crews who raided Japan on April 18, 1942, are at the museum. It was these land-based Doolittle Tokyo Raiders who flew off an aircraft carrier to give the Japanese an idea of the consequences they were to reap for Pearl Harbor. And Air Force bombers, fighters, and transports also flew to Europe. They gradually and determinedly stormed and pounded their way across the continent until the skies belonged to the Allies. Airpower could then concentrate its strength in the Pacific.

China and Burma were starting to benefit from the airlift that extended over the hazardous Himalayas, or the "Hump" route as it was known. United States forces had been island-hopping, but once again they were attacking the Japanese homeland, destroying industrial and military targets which supported the far-flung Japanese forces. Gen. Henry H. "Hap" Arnold's 20th Air Force, in particular, hammered aircraft production and oil supply centers. With Japan reeling from these attacks, Air Force B-29s delivered knockout blows—atomic bombs—on August 6 and 9, 1945. Japan on August 10 decided to surrender. With an invasion of the Japanese homeland no longer necessary, millions of American and Japanese lives had been spared from a long ground campaign that would have followed.

The second and final atomic bomb dropped on Japan fell from a Boeing B-29 Superfortress named the *Bockscar*. It is on display in its entirety at the Air Force Museum, along with casings of Little Boy and Fat Man, the two bombs that ended World War II.

Long before the war ended, President Franklin D. Roosevelt had directed that the U.S. Strategic Bombing Survey be established to study the results of United States bombing in Europe. Briefly stated, the Survey concluded: "Allied airpower was decisive in the war in western Europe. Hindsight inevitably suggests that it might have been employed differently or better in some respects. Nevertheless, it was decisive. In the air, its victory was complete; at sea, its contribution, combined with naval power, brought an end to the enemy's greatest naval threat—the U-boat; on land it helped turn the tide overwhelmingly in favor of Allied ground troops...."

In mid-August 1945, President Truman requested a similar study of the air war against Japan. The Pacific Survey concluded: "The experience of the Pacific war supports the findings of the Survey in Europe that heavy, sustained and accurate attack against carefully selected targets is required to produce decisive results when attacking an enemy's sustaining resources. It further supports the findings in Germany that no nation can long survive the free exploitation of air weapons over its homeland. For the future it is important fully to grasp the fact that enemy planes enjoying control of the sky over one's head can be as disastrous to one's country as its occupation by physical invasion." Findings of the two surveys have been reemphasized many times since then.

Visitors now continue their sojourn through those portions of the main museum complex displaying planes, habitats, equipment, photographs, and personal memorabilia from WW II. In addition to items displayed indoors, more of WW II is represented outdoors by the Eighth Air Force control tower, the two Nissen Huts, and monuments throughout the Memorial Park.

One of the four Curtiss P-36 Hawks to get airborne when the Japanese attacked Pearl Harbor was flown by 1st Lt. Philip M. Rasmussen, portrayed here climbing into his P-36A with a gun belt over his red-and-white striped pajamas. He had just awakened that Sunday morning when he looked out his dormitory window at Wheelus Field to see Japanese planes dropping bombs. Large scale destruction of aircraft and airfields made launching a major counter-attack impossible. But Lieutenant Rasmussen did manage to take off and shoot down one enemy aircraft. He went on to fly combat throughout the Pacific and shot down one more. In 1998, retired Colonel Rasmussen lectured at the museum.

The Japanese attack on Pearl Harbor on December 7, 1941, pushed the United States into World War II. This section of the museum features the nation's recruiting and training efforts on behalf of its military forces.

One of the first heroes of the war was Capt. Colin P. Kelly, Jr. This oil painting includes his B-17 bomber.

Newspaper headlines carried a series of defeats for the United States and its Allies as Japanese forces swept through the Southwest Pacific in the early months of 1942. When they attacked the Philippines, it was defended by only seventy-two American P-40Es, fifty-two obsolete P-35As, and twelve obsolete P-26s. The P-35s had been built for export to Sweden and some still carried Swedish markings when pressed into service.

Pilots received a portion of their aerial training in the earthbound Link trainer. Here they could safely practice instrument-flying techniques.

The air war against the Japanese and the Nazis is graphically presented in the main World War II display area. Historical photographs and maps, memorabilia donated by survivors and relatives, and crisp captions relate the story of the American struggle to restore freedom. A wing of the gigantic B-36 looms overhead.

American campaigns in Europe and Africa are portrayed along the right wall. The exploits of Ohioan Don Gentile are recognized in the vertical glass case. He was one of the Army Air Force's leading aces with 27.8 enemy planes destroyed. Gen. Dwight D. Eisenhower called him a "one man Air Force."

Flaps extended to assist in its takeoff, the first B-25 starts to lift from the aircraft carrier Hornet with Jimmy Doolittle at the controls. This raid in 1942 warned the Japanese that their homeland was not free from attack.

More than 1,000 women—1,074 to be exact—earned their wings as WASPs—Women's Air Force Service Pilots—between 1942 and late 1944 under the direction of Jacqueline Cochran. They freed men for combat as they performed a variety of missions. They logged sixty million miles, flight testing new aircraft, ferrying fighters and bombers to embarkation fields, towing targets for combat pilots and ground batteries to shoot at, and training other pilots in instrument navigation. Thirty-eight were killed serving their nation. Some of their contributions, as well as uniforms, are displayed.

These WASP pilots took B-17 transition training at Lockbourne, Ohio. "Pistol Packin' Mama" is painted on the nose of the Flying Fortress. *U.S. Air Force photo.*

As they did prior to World War I, American volunteers flew with their Allies prior to the Second World War. These panels and cases illustrate their contributions to the Chinese through the Flying Tigers and to the British through the Eagle Squadrons. Nazi war art also is included.

This device controlled the altitude of fire balloons, which the Japanese launched to drift across the Pacific Ocean to harass the United States. Fuses and black powder charges circled the frame.

Although the Curtiss-Wright AT-9 was officially known as the Fledging, it was more widely known as the "Jeep." Not easy to fly or land, this advanced trainer bridged the gap between single-engine trainers and twin-engine combat aircraft such as the Martin B-26 and Lockheed P-38. This AT-9 was built from parts of two, plus more parts built from scratch by museum restoration specialists.

To conserve scarce metals needed for combat aircraft, Beech built the airframe of its Beechcraft AT-10 Wichita out of plywood with only the engine cowlings and cockpit enclosure made of aluminum. This advanced trainer had superior performance, and more than half of the AAF's pilots transitioned from single-engine to multi-engine planes in them. Here a WASP is shown climbing into the cockpit.

Several trainers are seen in this view from the protocol lounge. Suspended are a Stearman PT-13D Kaydet (*left*) and a Fairchild PT-19A Cornell. Below them are a Curtiss-Wright AT-9 Fledging, Beech AT-11 Kansan, and a Curtiss P-40E Warhawk. The latter type aircraft is noted for its combat role in China with the Flying Tigers.

An instructor (*far left*) reminds a harried student pilot to keep the tail of the plane down when taxiing with the wind from behind. North American built this BT-9B Yale, also known as the NA-64A. Pilot training in the U.S. accelerated greatly in 1942 after the nation was brought into the war.

The British Spitfire interceptor was one of the most famous airplanes of WW II, with 20,351 built. Another 2,408 were built as Seafires for use from aircraft carriers. This Supermarine Spitfire Mark XI is a photographic reconnaissance version without guns or armor for increased speed. It is painted and marked as one that flew with the U.S. Eighth Air Force at Mount Farm airfield in England. Overhead are a Stinson L-5 Sentinel (*left*) and a Vultee L-1A Vigilant.

This Messerschmitt Bf 109G-10 is painted to represent an aircraft of the fighter unit that defended Germany against Allied bombers. Between 1936 and the end of WW II, some 33,000 Bf 109s were built for the Luftwaffe's use over Europe, North Africa, and the Russian Front. The one on display in 1999 replaced a similar appearing Hispano HA-1112-MIL that had been built in Spain after the war. It is nestled along side of a Martin B-26G bomber. The nose of a Waco CG-4A glider looms in the background.

Before the United States entered WW II, Americans in the Royal Air Force flew the Hawker Hurricane MK IIa with the 71 Squadron, one of three Eagle Squadrons, from May to August 1941. The Hurricane may best be known for its performance during the Battle of Britain, which started in July 1940, when 527 Hurricanes and 321 Spitfires successfully countered Hitler's 2,700 aircraft to maintain air superiority in the skies over Great Britain. The diorama above includes a flight line shed and display cases of WW II memorabilia, as well as ample signage on the aviators and their accomplishments.

Macchi MC-200 *Saetta* (Lightning) aircraft such as this were used by the Italians against U.S. forces in North Africa and over Italy itself. This Macchi was abandoned in North Africa, captured by the British, and eventually restored for the museum in Italy by a team from Aermacchi, the original builder.

The German Junkers JU-88 Zerstörer was one of the most versatile airplanes of World War II. It was used in practically every kind of combat role, even as a "pilotless missile." It made its first flight on December 21, 1936, and hundreds were still in use in 1945. The airplane on display is a JU-88D-1, a long-range photo reconnaissance version. On July 22, 1943, the *Baksheesh* was flown to Cyprus by a defecting Rumanian Air Force pilot. It was then flown by U.S. pilots to Wright Field where it was tested extensively. It was given to the museum in 1960. *U. S. Air Force photo.*

Britain's best fighter, the Mark V Spitfire, lost its superior role when the Germans introduced the Focke-Wulf FW-190 to combat in September 1941. Two years later, powered by a longer and more powerful engine, the FW-190D took on U.S. bombers. More than twenty thousand FW-190s of all types were built. The FW-190D-9 on display was captured and brought to the United States for testing. On the floor to the right is a Ruhrstahl X-4 Air-to-Air Missile, designed to shoot down B-17s. One was test fired from a FW-190 late in the war with the German pilot guiding the rocket-powered missile through the use of two long wires.

One of the best all-around fighters in the Pacific was the Kawanishi N1K2-J *Shiden-Kai* (George 21), produced during the last year of World War II by the Japanese. Only 428 were built because of initial production problems and later shortages of parts resulting from B-29 raids on Japan. This George 21 was donated by the City of San Diego in 1959 through the co-operation of the Air Force Association.

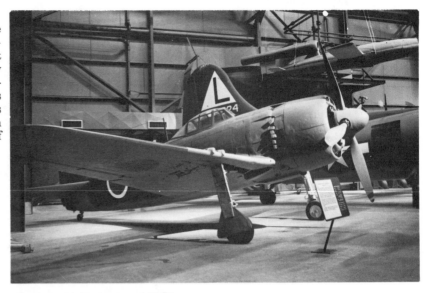

Developed from a 1938 design, the Messerschmitt Me-262 *Schwalbe* (Swallow) was the world's first operational turbojet aircraft. When initially flown as a pure jet on July 18, 1942, it proved to be much faster than conventional airplanes. More than 1,400 were built, but fewer than 300 saw combat. Allied bombers destroyed hundreds on the ground—much to the relief of other Allied aircrews. This *Schwalbe* was brought to the U.S. in 1945 for flight evaluation and was restored in 1976-79 at Kelly Air Force Base, Texas.

In light of the German dive bombing successes in Poland and France, the Army Air Corps acquired a small number of the U.S. Navy's SBD-3 Dauntless dive bombers which already were in production. Renamed and without a Navy tail hook, the Douglas A-24s saw limited action against the Japanese in 1942 in the Southwest Pacific. Some American crews referred to them as "Blue Rock Clay Pigeons." Many of the remaining A-24s finished their WW II careers in the U.S. as trainers or towing targets for gunnery practice.

Bell's P-39 Airacobra made its first flight in April 1939 at Wright Field and was one of the country's first-line pursuit craft by December 1941. Its engine (*left*) was located behind the cockpit, permitting a 37mm cannon to be fired through the propeller hub. Airacobras saw action throughout the world, particularly in the Southwest Pacific, Mediterranean, and Russian theaters. Of the 9,584 built, some 4,773 were given to the Soviet Union, whose pilots favored its ground-attack capability. The P-39Q on display is painted as the P-39J flown by Lt. Leslie Spoonts in 1942 during the Aleutian Campaign against the Japanese. On the wall behind the aircraft's tail are display panels that relate America's donation of 15,000 planes to the Soviet Union during WW II. They were ferried in mostly through Alaska and Iran.

Two views of Lockheed's P-38 Lightning designed in 1937 as a high-altitude interceptor. In Italy and the Mediterranean, New Guinea and the South Pacific, North Africa and the Aleutians, this two-engine, twin-tailed terror of the skies wrote aviation history. The P-38 was the most versatile fighter of World War II, and American flyers affectionately dubbed it the "round-trip ticket." One Lightning returned to base on a single engine five times. The Germans during the North African campaign named it the Forked-Tail Devil. *U.S. Air Force photo.*

Culver Aircraft Corporation was the major producer of radio-controlled target aircraft, such as this PQ-14B, during World War II. They were used to train anti-aircraft artillery gunners. A pilot occupied the cockpit only on ferry or check flights. Mr. Robert E. Parcell of Fort Worth donated this aircraft. *U.S. Air Force photo.*

According to legend the Japanese were afraid of sharks. To this end, World War II pilots painted their Curtiss P-40 Warhawks as shown in these three views. The legend has not been substantiated, but the Warhawk did have an outstanding record in downing 286 Japanese planes in an eight-month period, while only eight Warhawks were lost. The plane saw action on every fighting front of the war and was flown by the famed Flying Tigers as well as the first Army Air Forces black unit, the 99th Fighter Squadron. The airplane on display is one of the few P-40Es still in existence.

U.S. Air Force ph

The Republic P-47 Thunderbolt was one of America's leading fighter planes of World War II. An auxiliary fuel tank permitted the P-47 to escort heavy bombers far into German territory. Not only was it an impressive high-altitude escort fighter, the P-47 gained recognition as a low-level fighter-bomber because of its ability to absorb battle damage and keep flying. Republic Aviation Corporation donated the P-47D on exhibit.

U.S. Air Force photo.

Constructed primarily of plywood with a balsa wood core, the DeHavilland DH-98 Mosquito can be called an early stealth aircraft. Nearly 8,000 were built in Great Britain, Canada, and Australia. The famous British "Mossies" also were used by the U.S. AAF for photo and weather reconnaissance, and as night fighters. A mannequin is shown here painting black and white stripes on the wings shortly before the Normandy invasion so that the aircraft could be readily identified by Allied forces as a "friendly."

As if off on a mission, a Vultee L-1A Vigilant and a Stinson L-5 Sentinel appear over the heads of museum visitors. This L-1A, painted for ambulance service, was donated by Mrs. Lawrence Flahart of Anchorage, Alaska, in memory of her husband who had been rebuilding it. The L-5, painted as one used in New Guinea, was donated by Dr. Robert R. Kundel of Rice Lake, Wisconsin. Such airplanes were noted for their short field takeoffs and wide variety of support missions.

Groups of school children can be seen during the academic year on tours of the museum conducted by well-prepared volunteers from the Officers' Wives' Club at Wright-Patterson AFB. Tours for other visitors are announced on the museum's public address system.

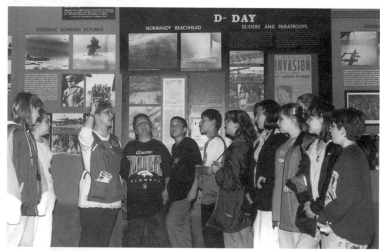

Dive bombing was to be its primary role, but North American A-36A Apaches also flew bomber escort and straffing missions in North Africa, Europe, and from India. This A-36A is painted as the one flown by Capt. Lawrence Dye in Tunisia, Sicily, and Italy. His *Margie H* is credited with twenty-five missions and two Nazi aircraft downed.

North American's P-51 Mustang could certainly be considered an international fighter. It was designed to British specifications by engineers formerly with Fokker and Messerschmitt. It gained fame with the U.S. Army Air Forces using a British engine. With the aid of external fuel tanks, it met Luftwaffe interceptors deep inside German territory and repeatedly scored heavily over the enemy planes. The P-51D on display was obtained from the West Virginia Air National Guard in 1957. It was the last USAF propeller-driven fighter in operation.

U.S. Air Force photo.

NORTHROP P-61C "BLACK WIDOW"

The heavily-armed Black Widow was this country's first aircraft specifically designed as a night fighter. In the nose, it carried radar equipment which enabled its crew of two or three to locate enemy aircraft in total darkness and fly into proper position to attack.

The XP-61 was flight-tested in 1942 and delivery of production aircraft began in late 1943. The P-61 flew its first operational intercept mission as a night fighter in Europe on July 3, 1944 and later was also used as a night intruder over enemy territory. In the Pacific, a Black Widow claimed its first "kill" on the night of July 6, 1944. As P-61s became available, they replaced interim Douglas P-70s in all US-AAF night fighter squadrons. During WWII, Northrop built approximately 700 P-61s; 41 of these were -Cs manufactured in the summer of 1945 offering greater speed and capable of operating at higher altitude. Northrop fabricated 36 more Black Widows in 1946 as F-15A unarmed photo-reconnaissance aircraft.

The Black Widow on display was presented to the Air Force Museum by the Tecumseh Council, Boy Scouts of America, Springfield, Ohio, in 1958. It is painted and marked as a P-61B assigned to the 550th Night Fighter Squadron serving in the Pacific in 1945.

SPECIFICATIONS

Span	66 ft.
Length	49 ft. 7 in.
Height	14 ft. 8 in.
Weight	35,855 lbs. loaded
Armament	Four .50-cal. machine guns in upper turret and four 20mm cannon in belly; 6,400 lbs. of bombs
Engines	Two Pratt & Whitney R-2800s of 2,100 hp. each

PERFORMANCE

Maximum speed	425 mph
Cruising speed	275 mph
Range	1,200 miles
Service ceiling	46,200 ft.
Cost	$170,000

Somewhere in France during World War II, armament men load 20mm shells into a box for the P-61 in the background. The shells were fired from four cannon in the belly of the Black Widow night fighter.

Russian and Free French pilots flew the Bell P-63 Kingcobra, courtesy of America's lend-lease program. Some of the aircraft were used by the U.S. for fighter training. This suspended P-63E was painted orange at the museum to resemble an RP-63A used for target practice. Such airplanes also had armor plating to protect them and their pilots from the .30 caliber lead-and-plastic bullets that disintegrated harmlessly upon contact with the Duralumin plating.

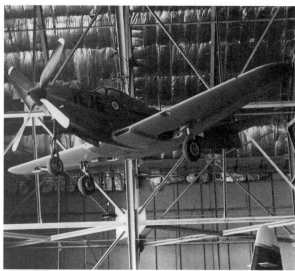

America's first jet-propelled airplane made its initial flight on October 1, 1942. Development of Bell's P-59 Airacomet was ordered personally by Gen. H. H. Arnold on September 4, 1941, and the project was conducted under the utmost secrecy. Bell produced sixty-six P-59s. Although the airplane's performance was not spectacular and it never got into combat, it provided training for AAF people and invaluable data for subsequent development of higher performance jet airplanes. A P-59B is exhibited.

The 1-A was America's first jet engine. Copied from the highly secret British Whittle jet engine, it had a thrust of only 1,650 pounds. The British provided the plans for the Whittle to America in 1941 and General Electric was requested to make a U.S. version. The work was conducted under such secrecy that many people employed on the project believed that they were building a turbosupercharger. At the same time Bell was requested to build an airplane that could use it. The result was America's first jet-propelled airplane: the P-59.

Similar in lines to the propeller-driven Bell P-63 King-cobra, the jet-powered Bell P-59A Airacomet never reached the operational stage of its "old-fashioned" sister. The P-63 was widely used by the Russian Air Force in support of ground troops. *U.S. Air Force photo.*

This prisoner of war display is the most extensive of the several on view at the museum. Many of the original items in the case are from Stalag Luft I, Barth, Germany, a Luftwaffe camp for captured American fliers.

Bandleader Maj. Glenn Miller disappeared on December 15, 1944, while on a flight from England to France. He was aboard a Noorduyn UC-64A Norseman, such as the one here (*top, center*). The exhibit honoring his band is located below the Canadian-built transport and behind the B-24D (*bottom*). To the right of the Norseman are a German Fieseler Fi 156 *Storch* and a Piper L-4 Grasshopper.

An enlarged exhibit honoring the Maj. Glenn Miller Army Air Forces Band was dedicated on July 3, 1976. Special guests for the event were Ray McKinley (*left*), who became leader of the famed group after Miller's untimely death in 1944, and Johnny Desmond, who was the lead vocalist for many years. "Next to letters from home, the Glenn Miller Army Air Forces Band was the greatest morale builder we had in the European Theater of Operations," reported Lt. Gen. James H. Doolittle. After the dedication, the current Glenn Miller Band performed for a crowd of 10,000 persons in a setting reminiscent of a bygone era.

Hanging from the ceiling of the large bay is this Taylorcraft L-2M Grasshopper. Such light planes were used in World War II for liaison and reconnaissance. An Aeronca L-3B Grasshopper, similar in appearance to the L-2M, is also part of the museum collection.

During World War II the Curtiss O-52 Owl was used for courier missions within the United States and for short-range submarine patrol over the Gulf of Mexico as well as the Atlantic and Pacific oceans. The plane on display was received from the U.S. Federal Reformatory at Chillicothe, Ohio, in 1962. Inside, painted on a bulkhead in the cockpit was the statement, "Help, I'm being held a prisoner."

The aircraft on display is one of 432 Expeditors that Beech rebuilt as C-45Hs. The C-45 was the World War II military version of the popular Beechcraft Model 18 commercial light transport. Beech built a total of 4,526 of these aircraft for the AAF between 1939 and 1945 in four versions: the AT-7 Navigator navigation trainer, the AT-11 Kansas bombing-gunnery trainer, the C-45 Expeditor utility transport, and the F-2 for aerial photography and mapping.

Student bombardiers, working in the nose of the Beech AT-11 Kansas, normally dropped 100-pound bombs filled with sand. About 90 percent of the Air Corps's 45,000 bombardiers were trained in AT-11s during World War II. This aircraft was donated to the museum by the Abrams Aerial Survey Corporation of Lansing, Michigan. *U.S. Air Force photo*.

U.S. Air Force photo.

The Curtiss C-46 Commando gained fame during World War II transporting cargo over the "Hump," the Himalayas, from India to China after the Japanese had occupied Burma. C-46s saw additional service during the Korean conflict. The C-46D on display was retired from service in Panama in 1968, stored at Davis-Monthan AFB in Arizona, and flown to the museum in May 1972. Suspended over the C-46D is a Cessna UC-78B Bobcat, which normally could carry one pilot and four passengers. Earlier models were the AT-8 and the AT-17.

Designed by the Waco Aircraft Co. of Troy, Ohio, this Waco CG-4A glider was made by the Gibson Refrigerator Co. in Greenville, Michigan. Hundreds were towed across the English Channel for the D-Day invasion of France on June 6, 1944. Each could carry thirteen troops or either a jeep, quarter-ton truck, or 75mm howitzer. Waco gliders also were used earlier in the Allied invasion of Sicily and in the China-India-Burma Theater. Fifteen companies made more than 12,000 CG-4As.

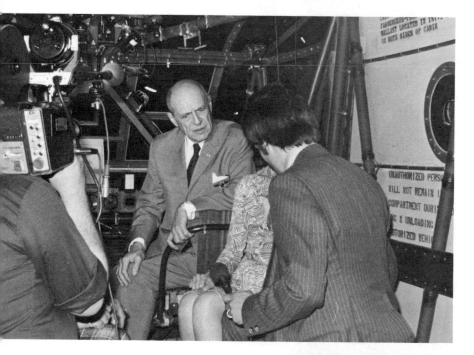

Gen. Matthew B. Ridgway (*center*) commanded the 82nd Airborne Division during World War II and used the Waco CG-4A glider during the invasion of Sicily in 1943 and Normandy a year later. The Army general participated in the 1976 unveiling ceremony for the glider at the Air Force Museum. He is shown here in the cockpit being interviewed for television.

Student glider pilots normally received about six hours of dual instruction in the Schweizer TG-3A training glider before being trained in the large CG-4A cargo glider. The TG-3A's wing is made of spruce and mahogany plywood covered with doped fabric. Henry A. Shevchuk of Cadogan, Pennsylvania, donated this glider to the museum. It was restored by the Sparton School of Aeronautics, Tulsa, Oklahoma.

Nearly 2,900 of these Fieseler Fi 156-1 Storch (Stork) aircraft were built for the German Air Force between 1937 and 1945. One of these light aircraft rescued deposed dictator Benito Mussolini from a rock-strewn Italian plateau. The airplane on display is painted as the one used by Field Marshal Erwin Rommel in North Africa. It was donated to the museum by Lt. Col. Perry A. Schreffler and Maj. Robert C. Van Ausdell, both of the U.S. Air Force Reserve.

This is the engine that powered the "Vengeance Weapon" developed by Germany during the war and fired against London and other city populations. Approximately 6,500 were manufactured during 1944/45. They had a maximum range of 220 miles, ceiling of 55 miles, and a speed of 3,500 miles per hour. Liquid oxygen and alcohol were used as propellants.

Visitors to the museum relive the news of the invasion of Europe and the eventual downfall of Nazi Germany.

After two years of testing and development, the Sikorsky R-4 Hoverfly was first used in combat in May 1944. The world's first production helicopter had shown such promise that the AAF ordered 100 R-4Bs. The one hanging from the ceiling of the museum was donated in 1967 by the University of Illinois. It has three rotor blades.

Skytrain is its official name, but the C-47 is more popularly known as the Gooney Bird. It was adapted from the DC-3, which appeared in 1936. During World War II it was used to carry troops and cargo, and to tow gliders. During the Vietnam conflict, the Gooney Bird was outfitted with machine guns to become Puff the Magic Dragon.

Six American crews and six British crews flew twelve Douglas A-20 Havoc planes on the first U.S. daylight bombing raid in Europe. It was a low-altitude mission against four Dutch airfields used by the Germans. The versatile A-20 attack bomber was used in the Pacific, Middle East, North African, Russian, and European theaters. The A-20G on display was donated by the Bankers Life and Casualty Company of Chicago. It was the first series to have a "solid" nose.

Coconut trees sway in the background as life-size mannequins appear to work on the museum's A-20G in New Guinea in 1944. The audio soundtrack for this habitat exhibit earned the 2001 Gold Visual Communications Award from the Ohio Museum Association. *U.S. Air Force photo by Jeff Fisher.*

U.S. Air Force photo.

Consolidated's B-24 Liberator was employed in every combat theater during the war. B-24s conducted one of the most famous raids of the war on August 1, 1943, against the oil refinery complex at Ploesti, Rumania. The raid set a record which still stands for the most Medals of Honor awarded in a single battle—five. The B-24D on exhibit is the same type airplane as the *Lady Be Good*, the world-famous B-24D which disappeared on a mission from North Africa in April 1943 and which was found in the Libyan Desert in May 1959. Volunteer museum tour guides who escort school groups refer to this plane as the *Strawberry Lady*.

U.S Air Force photo.

The Air Corps ordered 1,131 Martin B-26 Marauders in September 1940 because the plane had such good speed, range, and ceiling. Bombing from medium altitudes of 10,000 to 15,000 feet, the Marauder had the lowest loss rate of any Allied bomber— less than one-half of one percent. In 1945 when B-26 production was halted, 5,266 had been built. The B-26G on display was flown in combat by the Free French during the final months of World War II. It was obtained from the Air France training school in 1965.

On a simulated deck of the aircraft carrier USS *Hornet*, Lt. Col. James H. "Jimmy" Doolittle (*right*) discusses preparations with the ship's commander, Capt. Marc A. Mitscher, for the famous April 18, 1942, retaliatory attack on Japan. Built around a North American B-25B Mitchell, this diorama was created by the museum staff to make visitors feel they are back in 1942. More than 5,000 hours went into researching, planning, and building this habitat to honor the 80 men who risked their lives flying 16 B-25 bombers off the *Hornet's* deck for the daring raid over the islands of Japan. Now we know them as the Doolittle Tokyo Raiders. Several survivors of the raid and of the *Hornet's* crew attended the 1997 dedication of this expanded B-25 exhibit. The airplane on display was rebuilt by North American as one used on the Tokyo Raid. It was flown to the museum in April 1958 with then General Doolittle at the controls for a portion of the flight.

Nine mannequins in the diorama help reflect the human story behind the B-25 and the Tokyo Raid. Two crewmen (*above*) prepare ammunition while others (*right*) load bombs under the watchful eye of a carrier member. Also included is a six-minute video of a 1980 interview with General Doolittle recounting the Tokyo Raid.

First flown in 1939, the Douglas B-23 Dragon incorporated many features of the DC-3 commercial transport and was developed as a successor to the B-18. It was soon outclassed by more modern bombers, and the thirty-eight that were built were used in support roles. Several were still flying as private cargo transports in the early 1980s. *U.S. Air Force photo.*

The Douglas A-26 Invader, including the B-26, flew combat missions in three wars, over a twenty-six year span. It was used in World War II for level bombing, ground strafing, and rocket attacks. In 1948 it was redesignated the B-26 and used as a night intruder, harassing North Korean supply lines. During the Vietnam conflict it was used as a night interdiction aircraft along the Ho Chi Minh Trail. The A-26C on display appears in the colors and markings used during the Korean conflict.

U.S. Air Force photo.

One of the most famous airplanes ever conceived was designed by the Boeing Aircraft Company in 1934. The B-17 Flying Fortress made its maiden flight in 1935, in an era when practically all other bombers had two engines or less. Lt. Col. Robert D. Olds in February 1938 led six B-17s on a flight from Miami to Buenos Aires and set a new record for distance. One of the pilots was 1st Lt. Curtis E. LeMay, who ultimately became the Air Force chief of staff. The B-17 was used in every combat zone of World War II, particularly for daylight bombing of German industrial targets. Although only a handful of the Flying Fortresses were in service at the start of the war, nearly 13,000 had been constructed by May 1945 when production was halted. A B-17G is on display at the museum.

U.S. Air Force photo.

This exhibit (*below and left*) includes the tail of a B-17G that was shot down March 15, 1945, over Germany. It was recovered by the 52nd Equipment Maintenance Squadron in 1996 from a farmer's shed near the crash site. Tech. Sgt. Sator "Sandy" Sanchez, an aerial gunner as a volunteer on his third combat tour, died in the crash. Earlier a B-17 had been named the *Smilin' Sandy Sanchez*, the only known Flying Fortress named for an enlisted man. His photograph and seven medals are in the exhibit (*lower right*). The upper machine gun turret (*below, far left*) is similar to the type he used at times.

111

Photo by Jerry Rep.

This Boeing B-29 Superfortress, on display at the Air Force Museum, brought World War II to an end and halted the further loss of American lives. It is this aircraft from which the second atomic bomb was dropped on Japan. When Nagasaki was devastated on August 9, 1945, the Japanese government was finally convinced it must surrender. Overhead are a Sikorsky R-4B Hoverfly and the nose of Beech UC-43 Traveler.

From Salt Lake City, where the *Bockscar* crew had trained, to Nagasaki, where they demonstrated the proficiency of their training. That was the route of this famous B-29. It was named for Capt. Frederick C. Bock, the pilot, who switched planes and flew the *Great Artiste*, an instrument plane, for the eventual raid.

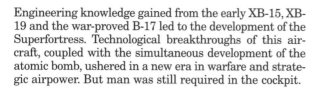

Engineering knowledge gained from the early XB-15, XB-19 and the war-proved B-17 led to the development of the Superfortress. Technological breakthroughs of this aircraft, coupled with the simultaneous development of the atomic bomb, ushered in a new era in warfare and strategic airpower. But man was still required in the cockpit.

Gen. Douglas MacArthur (*large photograph*) accepted the Japanese surrender aboard the USS *Missouri* on September 20, 1945, in Tokyo Bay. The panel left of center includes a profile photograph of the B-29 from which the final atomic bomb was dropped. That aircraft now is displayed in the museum.

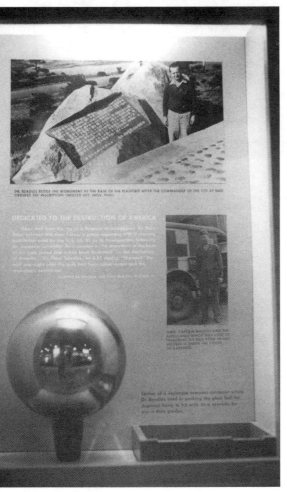

"Dedicated to the Destruction of America" is what was proclaimed on a monument at the base of a Japanese flag pole at Irumagawa (later Johnson) AB near Tokyo. When the base was occupied by the Army Air Forces, the monument was destroyed and the gold-colored ball (*bottom left*) that had topped the pole was "liberated" one night by Dr. Elmer Beadles, an AAF dentist, shown twice in the display panel.

Although the war had ended, not all Americans returned home. The B-24 *Lady Be Good* crew members perished in the African desert in 1943, but their fate was not discovered until 1959. Their courage is depicted here and in the Chapel exhibit.

Depending on engines and interior configuration, these Lockheed C-60A Lodestars also were known as C-56s, C-57s, and C-59s. During WW II they were used for training and for transporting personnel and freight. After the war they were sold to private operators for use as cargo or executive transports. The C-82A is in the background.

One prototype of the Fairchild C-82A Packet flew in 1944 before operational deliveries began in late 1945. Its primary use was for large cargo and personnel hauling, and often for paratroop operations. This Packet has red arctic markings for high visibility.

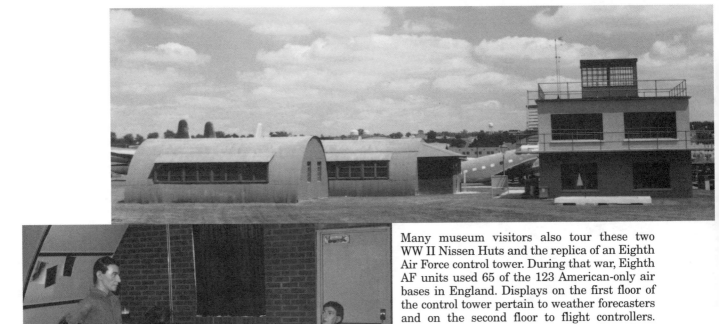

Many museum visitors also tour these two WW II Nissen Huts and the replica of an Eighth Air Force control tower. During that war, Eighth AF units used 65 of the 123 American-only air bases in England. Displays on the first floor of the control tower pertain to weather forecasters and on the second floor to flight controllers. One of the two Nissen Huts shows how these quonset huts were used to brief crews before missions and the other where they might have relaxed afterwards. Museum volunteers in both locations are available to assist visitors.

★ 6 | *The Jet Age*

After victory in World War II, the importance of airmen and their machines was proved beyond doubt. The Jet Age had arrived and it was clear that future defense would depend largely on technological superiority. In October 1942 the XP-59A jet aircraft made its first flight. Development of this revolutionary aircraft was shrouded in secrecy, to the point of placing a wooden propeller on the nose to confuse enemy espionage efforts.

With knowledge gained from the XP-59A, a version of which is on display at the Air Force Museum, development of the Lockheed F-80 Shooting Star began in 1942. But operational models had reached the Air Force too late for WW II. Nevertheless, the F-80 is assured a lasting place in aviation history as the first jet accepted for operational service within the Air Force. With it, the Jet Age became a reality for America. The Shooting Star on display at the museum was especially modified for racing by equipping it with a smaller canopy, a shorter wing, and redesigned air intakes.

Perhaps the world's most exotic jet airplane is the North American XB-70 Valkyrie, which continues to dominate the display of experimental aircraft now across the field in Hangar 9. This six-engine jet bomber was conceived in the 1950s to fly at three times the speed of sound. Because of fund limitations, only two were built, both for the advance study of aerodynamics and propulsion as related to large supersonic aircraft. The Valkyrie on display is the only surviving XB-70. It flew as a research ship from September 1964 until it was flown to the museum in February 1969. Scientific data gathered over the years from the XB-70 have gone into the development of other jet aircraft, both military and civilian, including the supersonic Concorde.

On July 26, 1947, President Truman signed the legislation that established a new defense organization for the nation and brought an independent Air Force into being. This National Security Act of 1947, appropriately, was signed aboard

U.S. Air Force photo.

America's first jet airplane manufactured in large quantities was the Lockheed P-80 Shooting Star, which was redesignated the F-80 in 1948. It made its initial flight in January 1944, but was produced too late for World War II. During the Korean conflict it was used extensively for low-level attacks against ground targets. Its basic design was used for the T-33 and to a lesser degree for the F-94. The P-80R on display was modified for racing and set a world speed record in 1947 of 623.8 miles per hour.

the President's personal airplane, the "Sacred Cow," now on display at the museum. The Air Force began functioning as the nation's specialist in airpower on September 18 of that year when W. Stuart Symington was sworn in as its first secretary.

A year later, the new Air Force saw its first action under its new name. It wasn't combat. It was Operation Vittles, the airlift to save the people of West Berlin from Communist-imposed starvation. Since then, Air Force disaster relief teams have contributed to the welfare of people in many other parts of the world as well as the United States. And the Air Force has aided the civilian populace, at least indirectly, through numerous by-products of its research, development, and training.

When Communist aggression erupted in Korea in 1950, the Air Force was quick to respond. As the ground battle roared back and forth, the Air Force once again proved the value of hitting the enemy behind the lines, providing close tactical support for ground forces and air supply for isolated land operations. It was also an air war as jet fighters edged their way into the skies and propeller-driven aircraft began slipping into the background. The struggle also produced thirty-eight Air Force aces and four Medal of Honor recipients who are commemorated at the museum.

In the 1960s the Air Force was again called upon to provide close air support to ground forces, this time in Southeast Asia. American air action over South Vietnam continued from late 1961 to early 1973. In the North, it was exclusively an air war. Air Force fighter-bombers and bombers attacked military barracks, storage areas, infiltration routes, and lines of communication. Because of the longstanding White House policy of graduated escalation, our airmen were forced to encounter the most sophisticated and concentrated air defense network ever faced in any war. And while the United States was never denied use of enemy airspace, the Air Force did not gain the clear-cut air supremacy it enjoyed in WW II and Korea. Nevertheless, American ground troops never knew the demoralization of having their source of supplies cut off, of being constantly exposed to aerial reconnaissance, or of always being vulnerable to strafing and bombing. Finally, airpower forced an end to the long American involvement in Southeast Asia. Planes on display at the museum from this era include the actual A-1E which Maj. "Bernie" Fisher flew on the mission that earned him the first Medal of Honor awarded to an airman in the Southeast Asia conflict.

Jet-propelled aircraft were the mainstays of that conflict, but propeller-driven aircraft played their important support role, as they will continue to in the future. The success of all flying operations, however, remains with the vision, experience, courage, and dedication of the people who command, support, and fly them. This was demonstrated vividly in early 1991 against Iraqi forces during Operation Desert Storm. That campaign was summarized by the U.S. Air Force chief of staff as "the first time in history that a field army has been defeated by airpower." And as the decade came to a close, the Air Force continued to serve our nation in peacekeeping and humanitarian missions around the world. Then on September 11, 2001, that service was again escalated as Air Force people and equipment responded against a new brand of Terrorism.

The aircraft, men, and women who have contributed over the years to jet-age history are honored throughout most of the main museum complex and also across the field in Hangars 1 and 9, the Presidential Aircraft Hangar and the Research and Development/Flight Test Hangar.

During the new century, Air Force pilots are to fly the F-22 air superiority fighter at supersonic speeds while merely in cruise mode. A prototype version of the Lockheed-Boeing-General Dynamics fighter, the YF-22, was introduced in the Modern Flight Hangar in 1997 at the time of the first flight of the production model. Overhead is a Beech T-34A Mentor, used in the 1950s for primary flight training with a cruising speed of 173 mph.

President Truman signed the legislation that created an independent Air Force, aboard his Douglas C-54, popularly known as the "Sacred Cow." The Air Force now celebrates its birthday on September 18, the date in 1947 when W. Stuart Symington became its first secretary.

When the Soviet Union attempted to starve and freeze the people of West Berlin, an American and British airlift delivered more than 2.3 million tons of food, fuel, and other supplies from June 1948 to July 1949. During that airlift, Lt. Gail S. Halvorsen started Operation Little Vittles by dropping candy in toy parachutes (*left panel, above*). U.S. Air Force photo.

President Truman in 1948 signed an Executive Order that banned all forms of segregation in the military. A copy of that two-page decree is displayed alongside photographs of two prominent Tuskegee airmen: Lt. Gen. Benjamin O. Davis and Gen. Daniel J. "Chappie" James, Jr.

The Air Force's first postwar fighter, the Republic F-84 Thunderjet, first flew on February 28, 1946. From 1947 to 1953 approximately 4,450 were built with "straight wings." They were used in the Korean conflict against enemy railroads, bridges, supply depots, and troop concentrations. The F-84E on display is nestled alongside a B-36.

REPUBLIC F-84F "THUNDERSTREAK"

The swept-wing F-84F evolved from the straight-wing F-84. The prototype first flew on June 3, 1950 and deliveries began in 1954, primarily to the Tactical Air Command as a ground support fighter-bomber.

Republic produced 2,112 -Fs while General Motors built an additional 599. Of these, 1,301 were to NATO air forces. Production of a reconnaissance version, the RF-84F, totaled 718 aircraft, including 386 for allied countries. The RF-84F featured engine air intakes at the wing roots plus cameras in the nose.

F-84Fs gradually were replaced by supersonic F-100s in the late 1950s and were turned over to Air National Guard units. However, some F-84Fs were called back to temporary USAF service in the early 1960s due to the Berlin Crisis and the threat of Soviet missiles in Cuba.

The aircraft on display was flown to the museum in 1970 following its assignment to the Ohio ANG. During its career, it served in England, Greece, Alaska, and the continental U.S. It was one of the aircraft which participated in the mass deployment of fighters across the Atlantic in November 1961.

SPECIFICATIONS

Span	33 ft. 7 in.
Length	43 ft. 5 in.
Height	15 ft.
Weight	27,000 lbs. loaded
Armament	six .50-cal. machine guns and 24 five-inch rockets; 6,000 lbs. of bombs
Engine	Wright J65 of 7,220 lbs. thrust

PERFORMANCE

Maximum speed	685 mph
Cruising speed	535 mph
Range	1,900 miles
Service ceiling	44,450 ft.
Cost	$769,000

Republic's YRF-84F version of the Thunderstreak was designed to hitch a ride on the underside of a B-36 and tag along (*see hook on nose*) until needed for protection or photo reconnaissance. The concept of the 1950s was dropped with the development of aerial refueling. Overhead is the Boeing YQM-94A Compass Cope B. The remotely piloted vehicle was tested in the mid-1970s on high altitude, long-endurance photo reconnaissance and electronic surveillance missions.

LOCKHEED F-94A "STARFIRE"

The two-place F-94 was this nation's first operational jet all-weather interceptor. It was developed from the single-seat P-80 Shooting Star which had been the Army Air Forces' first operational jet aircraft procured in significant quantities. Although the F-94 had a redesigned fuselage, it used the P-80 tail, wing, and landing gear. The Starfire was also the first U.S. production jet to have an afterburner, which provided brief periods of additional engine thrust. It was equipped with radar in the nose to permit the observer in the rear seat to locate an enemy aircraft at night or in poor weather. The pilot then flew the Starfire into proper position for an attack based upon the observer's radar indications.

F-94s were primarily deployed for the defense of the United States in the early 1950's serving with Air Defense Command squadrons. Many Air National Guard units were later equipped with F-94s.

Lockheed produced 853 F-94s for the Air Force, beginning in December 1949. Of these, 110 were F-94As. The F-94A on exhibit was transferred from active inventory to the Air Force Museum in May 1957.

SPECIFICATIONS

Span . . . . 38 ft. 9 in.
Length . . . . 40 ft. 1 in.
Height . . . . 12 ft. 2 in.
Weight . . . . 15,330 lbs.
Armament . . Four .50-caliber machine guns.
Engine . . . . . General Electric J33 of 6,000 lbs. thrust with afterburner

PERFORMANCE

Max. speed . . . . . . 630 mph.
Cruising speed . . . . 520 mph.
Normal range . . . . 930 miles
Service ceiling . . . 42,750 ft.

Developed by North American in 1945, the F-82 Twin Mustang was produced by "marrying" two P-51 airframes. The purpose was to provide a fighter that would have two pilots, thereby reducing fatigue on long-range escort missions to Japan. However, World War II ended before the plane went into production. With the advent of the Korean conflict, the F-82 was used as a night fighter. The first three North Korean airplanes destroyed by U.S. forces were shot down by all-weather F-82 interceptors on June 27, 1950. The F-82B on display *(front and rear view)* set a new distance record in 1947 when it was flown nonstop 5,051 miles from Hawaii to New York.

Three models of the North American F-86 Sabre are displayed at the museum, one in a unique manner—without its skin. The F-86H, with its anatomy revealed, was a fighter-bomber. The F-86A, shown with the control tower sign from Japan's Itazuke Air Base, was designed as a high-altitude day fighter. The F-86D, an all-weather interceptor, is shown in the top photograph. Its noticeable characteristic is a black nose cone containing radar and a wider air scoop beneath. During the Korean conflict, the F-86A, E, and F models shot down 829 Russian-built MIG-15s at a loss of only 58 Sabres.

NORTHROP F-89J "SCORPION"

The F-89 was a twin-engine, all-weather fighter-interceptor designed to locate, intercept, and destroy enemy aircraft by day or night under all types of weather conditions. It carried a pilot in the forward cockpit and a radar operator in the rear who guided the pilot into the proper attack position. The F-89 made its initial flight in Aug. 1948 and deliveries to the Air Force began in July 1950. Northrop produced 1,050 F-89s.

On July 19, 1957, a Genie test rocket was fired from an F-89J, the first time in history that an air-to-air rocket with a nuclear warhead was launched and detonated. 350 F-89s were converted to "J" models which became the Air Defense Command's first fighter-interceptors modified to carry nuclear armament.

The Scorpion on display was transferred to the Air Force Museum from the Maine Air National Guard and was the last F-89 in service with an operational unit.

SPECIFICATIONS

Span . . . . 60 ft. 0 in.

Length . . . . 53 ft. 10 in.

Height . . . . 17 ft. 6 in.

Weight . . . . 47,719 lbs

Armament . Two AIR-2A Genie missiles plus four AIM-4C Falcon missiles

Engines . . Two Allison J35 jet engines of 8,000 lbs. thrust each with afterburner

PERFORMANCE

Maximum speed . 630 mph.

Cruising speed . . . 465 mph.

Range . Approx. 1,000 miles

Service ceiling . . . 45,000 ft.

When the F-89J was displayed outdoors, Genie atomic rockets were hung under its wings. Framed within the tail fins of this rocket is the Wright Field laboratory that once housed an Air Force nuclear reactor. Scientists there unsuccessfully tried to build an atomic engine for aircraft.

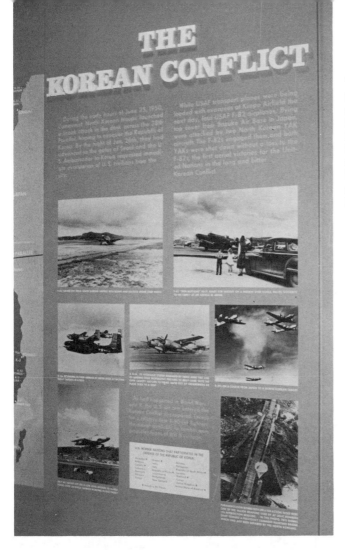

"During the early hours of June 25, 1950, Communist North Korean troops launched a sneak attack in the dark across the 38th parallel, hoping to conquer the Republic of Korea." Two days later, while four USAF F-82s were flying protective cover for American evacuees near Seoul, two North Korean YAK aircraft attacked the USAF fighters. Both of the enemy planes were shot down without an American loss. Thus began "the long and bitter Korean Conflict."

Many of the museum's exhibits, such as this one, were made possible by donations from individuals who contributed in some measure to aviation history. For instance, the original South Korean citation, *top right,* was donated by an Indianapolis resident.

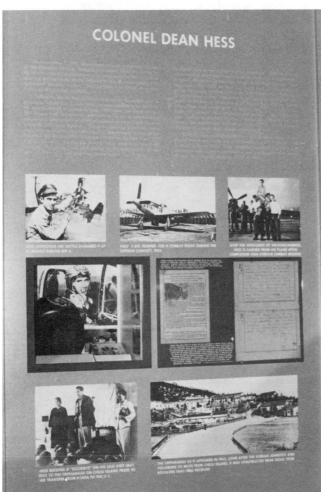

This exhibit recognizes the work of a former Ohio minister who served as a fighter pilot in World War II and Korea. However, Colonel Dean Hess is best noted for arranging an airlift to rescue thousands of Korean orphans from Communist armies sweeping in from the north. His "Operation Kiddy Car" flew the youngsters to safety on Cheju Island where he established an orphanage.

The North American F-100 Super Sabre made its combat debut in Vietnam, where it was used extensively as a fighter-bomber in such missions as attacking bridges, road junctions, and troop concentrations. Earlier, in 1953, an F-100 became the world's first operational aircraft known to be capable of supersonic speeds (with afterburner) in level or climbing flight. The F-100C shown here, was introduced in 1955 with an improved electronic bombing system. This Super Sabre was named the *Susan Constant* on May 12, 1957, by the wife of the Speaker of England's House of Commons for the 350th anniversary of the founding of Jamestown, Virginia. Overhead is a Cessna O-2A Skymaster used in Vietnam to mark enemy targets with smoke rockets for other aircraft to attack. In the background is the primary classroom of the museum's Education Division.

Visitors learn from the sign that this staff sergeant is filling the Super Sabre's liquid oxygen (LOX) system so the crew could breath oxygen at high altitudes. Such a system was introduced in the late 1950s to save weight and space over gaseous oxygen systems then in use. The bent pipe over the mannequin's head is part of a probe that runs back into the right wing. It permitted in-flight refueling to extend the aircraft's 2,000-mile range.

A flattened nose is the hallmark of the RF-101, shown here landing at Tan Son Nhut Air Base in South Vietnam. This reconnaissance version took low-altitude photographs of Soviet missile sites during the 1962 Cuban crisis and of enemy activities during the late 1960s in Southeast Asia. *U.S. Air Force photo.*

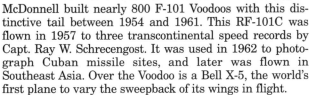

McDonnell built nearly 800 F-101 Voodoos with this distinctive tail between 1954 and 1961. This RF-101C was flown in 1957 to three transcontinental speed records by Capt. Ray W. Schrecengost. It was used in 1962 to photograph Cuban missile sites, and later was flown in Southeast Asia. Over the Voodoo is a Bell X-5, the world's first plane to vary the sweepback of its wings in flight.

The Convair F-102A Delta Dagger on display was one of the first Air Force planes to intercept and convoy a Russian TU-20 Bear long-range bomber over the Arctic. At the time it served with a fighter-interceptor squadron in Iceland. A Ryan Firebee drone, used for target practice, appears to be attacking the F-102A. This Firebee set a record of twenty-fire flights from 1958 to 1960. The radio-controlled drone deployed its own parachute after each mission so that it could be reused.

This photograph was taken during the September 1968 NATO fleet exercises while the Russian Bear was being tracked near the coast of Iceland. Other Soviet aircraft intercepted by F-102s that year included the TU-16 Badger, TU-95 Bear, and M-4 Bison. *U.S. Air Force photo.*

Because of its high landing speed, the F-102A would release a parachute from the compartment above its exhaust section as a breaking device. The F-102 was the world's first supersonic all-weather jet interceptor.

Lockheed's F-104 Starfighter was the first aircraft to hold simultaneous world records for speed, altitude, and time-to-climb. The F-104C on display served with the U.S. Air Force in West Germany, Spain, Taiwan, Vietnam, Laos, and Thailand. In 1962 it won first place in the "William Tell" Fighter Weapons Meet.

Mannequins dressed as fire fighters in this life-size diorama go to the aid of the "Thud" pilot. This particular Republic F-105G Thunderchief is credited with downing three MiGs over Vietnam. It is painted as when Americans flew it from Korat Royal Thai Air Base in 1972-73. The "Thud" was developed as a supersonic tactical fighter-bomber to replace the F-84F during the 1950s.

This Convair F-106A Delta Dart, which was developed from the F-102 Delta Dagger, landed itself on a snow-covered field after the pilot had been forced to eject because of a malfunction. Such all-weather interceptors first flew on December 26, 1956 and became operational with the Air Defense Command in July 1959. *U.S. Air Force photo.*

The North American F-107A made its initial flight on September 10, 1956 reaching Mach 1.03. It never went into production, losing out to the F-105 as the standard fighter-bomber for the Tactical Air Command.

127

Although it first flew in August 1955, use of the Lockheed U-2 was kept secret until May 1960 when a civilian-piloted U-2 on a reconnaissance flight over the Soviet Union was downed by a barrage of SAM-2 missiles at 70,000 feet. The one suspended at the museum, the last U-2A built, made 285 flights during the 1960s to gather data on clear air turbulence at high altitudes. In the 1970s it tested reconnaissance systems. It was delivered to the museum in May 1980. The tail of a LTV A-7D Corsair II is at the right.

Lockheed also produced the SR-71 for long-range, strategic-reconnaissance missions. From the 1960s through the 1980s these "Blackbirds" (as they are unofficially known) flew missions that remain classified. Then the Air Force retired its SR-71s in January 1990 because of a decreasing defense budget and high operating costs. A few were temporarily used later in the decade. Maj. (later General) Jerome F. O'Malley and Maj. Edward D. Payne made the first operational SR-71 sortie in this Blackbird on March 21, 1968. During its career, it flew 2,981 hours and 257 missions—more than any other SR-71.

Hundreds of the Northrop F-5 Freedom Fighters have been procured by the Air Force for use by allied nations around the world. They resemble the supersonic T-38 trainer. An American squadron flew combat missions with the YF-5A in Southeast Asia in 1966-67 to evaluate its combat performance. The YF-5A displayed was retired to the museum in 1970.

In spite of its size, the Cessna YA-37A is not a toy. The aircraft on display alongside the B-36 is one of two YAT-37D models. It was retired to the museum in December 1964 and recalled to active duty in August 1966 for final design testing of the A-37 attack model, then urgently needed for close air support of ground troops in South Vietnam. It was retired for a second time in July 1970 as the YA-37A.

Visitors easily peer into the Cessna YA-37A, which was modified from the T-37B primary trainer to evaluate it as a counterinsurgency attack and reconnaissance aircraft.

The twin-engined Cessna has two primary models: the T-37 trainer, seen over calm waters, and the A-37 Dragonfly, shown heavily loaded on a mission over Southeast Asia. One of the Cessna's features is its ability to carry out a mission with only one engine operating.

U.S. Air Force photo.

Although the conflict in Southeast Asia erupted during the "Jet Age," the first airman to win a Medal of Honor for service there was a pilot who flew this propeller-driven Douglas A-1E Skyraider. Maj. Bernard F. Fisher received the award from President Johnson on January 19, 1967.

Artist: Harvey Kidder

On March 10, 1966, Major Fisher had rescued a fellow pilot shot down over south Vietnam in the midst of enemy troops. A year later the repaired Skyraider was returned to the United States for preservation at the museum. The rescued pilot flew it from California to Ohio.

This Cessna 0-2A served with the 20th Tactical Air Support Squadron at Da Nang Air Base in the late 1960s. It is one of 346 special "push-and-pull" Cessna 337 Skymasters that entered U.S. Air Force service in 1966 to replace the 0-1 in the forward air controller role in Vietnam. *U.S. Air Force photo.*

Cessna O-1G Bird Dogs were used in Vietnam as two-place observation aircraft. Rockets were launched from the wing pods to mark enemy ground positions for other aircraft to strike. *U.S. Air Force photo.*

U.S. Air Force photo.

After WW II, air-sea rescue duties of the Air Rescue Service required special planes for overwater missions. So, in 1948, the Air Force acquired eight surplus Grumman J2F-6s from the Navy. These became OA-12 Ducks. Grumman originally sold this Duck in 1945 to the Coast Guard. Eventually it had a series of civilian owners and even a movie career, including the film Murphy's War in the early 1970s.

Designed to meet a Navy requirement, the Grumman HU-16 Albatross was able to operate from land or water and, with skis, from snow and ice. The prototype first flew on October 24, 1947, and was known as the SA-16A. During the Korean conflict the Albatross rescued nearly 1,000 United Nations personnel from coastal waters and rivers. The HU-16B on display was the last operational USAF Albatross. Two weeks before it was flown to the museum, it set a world altitude record for twin-engine amphibians when it reached 32,883 feet on July 4, 1973. Overhead is a Douglas O-46A.

An HU-16 on a search and rescue flight over Labrador in 1967. *U.S. Air Force photo.*

Several hundred downed fliers were rescued during World War II by crews flying the Consolidated OA-10 Catalina, the Army Air Force's version of the Navy PBY series seaplanes and amphibians. This aircraft was flown extensively by the Brazilian Air Force until 1981 in humanitarian roles in the Amazon Basin. It was flown to the museum three years later. *U.S. Air Force photo.*

The Sikorsky H-5 helicopter gained its greatest fame during the Korean Conflict when it repeatedly rescued United Nations pilots from behind enemy lines and evacuated wounded troops from frontline areas. Displayed is one of the twenty-six YH-5As ordered in 1944. More than 300 of various models were built by 1951. In the background is a small portion of the Air-Sea Rescue Exhibit.

Several fragile-appearing helicopters hang from ceilings in the museum. The closest here is the McDonnell XH-20 Little Henry, the world's first ram-jet helicopter, which flew in 1947-48. The jets are located on the ends of the two rotor blades.

In addition to serving with the Air Force, the Vertol H-21 Workhorse was supplied to the U.S. Army, the French Navy, the Royal Canadian Air Force, and the West German Air Force. The "Flying Banana" made its first flight in April 1952 and could carry twenty fully equipped troops or twelve litter patients. Displayed is a CH-21B.

One of President Eisenhower's original two helicopters is on display at the museum. He became the first U.S. President to fly in a helicopter when an Air Force Bell UH-13J Sioux, a sister ship of the Sioux on display, carried him from the White House lawn on July 13, 1957.

Bensen's X-25A Gyrocopter was tested in 1968/69 as a means for downed flyers to escape from enemy territory.

KAMAN HH-43B "HUSKIE"

The "Huskie" was used primarily for aircraft crash rescue and fire-fighting. Delivery of the first H-43As to the USAF began in November 1958 for assignment to Tactical Air Command bases. Delivery of the -B series began in June 1959; it had better performance and could lift a greater load. (In mid-1962, the USAF changed the H-43 designation to HH-43 to reflect the aircraft's rescue role.) HH-43Fs (the final USAF version) were used in Southeast Asia as "aerial fire trucks" and for rescuing downed airmen in both North and South Vietnam.

A Huskie on rescue alert could be airborne in approximately one minute. It carried two rescuemen/fire-fighters in its cabin and a fire suppression kit hanging beneath it. Foam from the kit plus the powerful downwash of air from the rotors was used to open a path to crash victims.

The HH-43B on display, one of approximately 175 -Bs purchased by the USAF, established seven world records in 1961-1962 for helicopters in its class for rate of climb, altitude, and distance traveled. It was assigned to rescue duty with Detachment 3, 42nd Aerospace Rescue and Recovery Squadron, Kirtland AFB, N. Mex., prior to its retirement and flight to the museum in April 1973.

SPECIFICATIONS

Rotor diameter	47 ft.
Overall length	47 ft.
Height	17 ft. 2 in.
Weight	9,150 lbs. loaded
Armament	none
Engine	Lycoming T-53 of 860 hp.

PERFORMANCE

Maximum speed	120 mph
Cruising speed	105 mph
Range	185 miles
Service ceiling	25,000 ft.
Cost	$304,000

It was first flown in 1949, but the Sikorsky H-19 saw service for several decades in the United States and overseas. For instance, it was used for rescue missions during the Korean Conflict and for search missions off of West New Guinea in 1962 for the United Nations.

These two aircraft supported President Johnson. The North American T-39A Sabreliner (*tail view*) was assigned for that purpose from 1968 to 1973 at Bergstrom AFB, Texas. The Beech VC-6A was used to transport the President and members of his family between Bergstrom AFB, near Austin, and the Johnson family ranch near Johnson City. At that time it was known informally as the "Lady Bird Special," named after Mrs. Johnson. Both aircraft were retired in the mid-1980s to the museum.

There are no national markings on this Sikorsky CH-3E helicopter. Painted flat black, it was nicknamed the "Black Maria" and flew highly classified special missions in Southeast Asia. Bullet holes are still visible on the inside. The HH-3E versions were known as "Jolly Green Giants" and were flown extensively in combat rescue missions.

During the 1960s, McDonnell Douglas F-4s set many speed, time-to-climb, and altitude records. The Phantom II carried a crew of two and three times the normal bomb load of a WW II B-17. It was flown by the Air Force in Vietnam on interception, close air support, and bombing missions. The F-4C on display was piloted by Col. Robin Olds on May 20, 1967, when he and Lt. Stephen B. Crocker (backseat pilot) shot down two MiG-21s. A WW II ace, Colonel Olds was the first USAF pilot to score four victories with F-4s in Southeast Asia. He flew out of Thailand.

The McDonnell Douglas F-15 Eagle was the first U.S. fighter to have engine thrust greater than the normal weight of the aircraft, allowing it to accelerate while in a vertical climb. During Operation Desert Storm, F-15Cs escorted strike aircraft over long distances and shot down thirty-six Iraqi planes. An A model is shown here. Overhead is a Canadian-built DeHavilland U-6A Beaver, first known as the L-20, used in the late 1950s and early 1960s in utility transport and liaison roles.

General Dynamics built more than 500 versions of the F-111 before production ended in 1976. This F model Aardvark is painted and marked as flown in early 1991 during Operation Desert Storm against Iraq. *Miss Liberty II* was flown to the museum in July 1996 when the F-111s were retired. These Mach 2.5 tactical bombers also were used against North Vietnamese forces. They have wings that were straight out for takeoffs, landings, and slow-speed flight, but swept back for high speeds.

One of Lockheed's first F-117A Stealth fighters was delivered to the museum in July 1991 with thousands of spectators on hand for close-up views of the once-secret aircraft. The museum's Stealth is a full-scale development model similar to operational craft used against Iraqi forces. During Operation Desert Storm, F-117As flew about 2.5 percent of the combat sorties, but hit more than 30 percent of the strategic targets without receiving combat damage. *U.S. Air Force photo.*

MIKOYAN-GUREVICH MiG-15 "FAGOT"

The MiG-15 was developed by the Soviet Union following WW II. It began appearing in service in 1949, and by 1952 it had been provided to various Communist satellite nations, including North Korea where it was used extensively against United Nations forces.

The airplane on display was flown to South Korea on September 21, 1953 by a defecting North Korean pilot who was given a reward of $100,000. The airplane was subsequently flight-tested on Okinawa and then brought to Wright-Patterson AFB for additional flight tests. An offer by the U.S. to return the airplane to its "rightful owners" was ignored and in November 1957, it was transferred to the Air Force Museum for public exhibition.

SPECIFICATIONS

Span . . . . 33 ft. 1-1/2 in.
Length . . . . .36 ft. 4 in.
Height . . . . .11 ft. 2 in.
Weight . . .11,270 lbs. loaded.
Armament . . . . . .Two 23mm cannon and one 37mm cannon, plus 2,000 lbs. of bombs or rockets.
Engine . . . .VK-1 of 6,000 lbs. thrust (copy of British Rolls-Royce "Nene" engine).

PERFORMANCE

Maximum speed . . 670 mph.
Cruising speed . . . .525 mph.
Range . . . . 500 miles.
Service ceiling . . . .51,000 ft.

Soviet-designed MiG-21 Fishbeds have been flown by some 30 air forces around the world. This fighter was acquired by the Air Force Museum in 1989 through a trade with an entrepreneur who had purchased it in Romania. The MiG-21 probably had been built in Czechoslovakia. A MiG-17 also displayed at the museum was obtained from Egypt. A Boeing B-47 is in the background.

137

President Franklin D. Roosevelt made his first and only flight in this Douglas VC-54C Skymaster when he traveled to Yalta in February 1945 for a conference with Soviet and British leaders. It was nicknamed the "Sacred Cow," probably by the news media. In 1947, aboard this transport, President Truman signed the legislation that created the independent Air Force. Overhead is a North American T-28B Trojan.

This Douglas VC-118 Liftmaster is a modified DC-6 commercial airliner that served President Truman starting July 4, 1947. It was then named the *Independence* for his hometown in Missouri. It was modified to include a state-room in the aft section. The main cabin could seat twenty-four passengers or be made up into twelve "sleeper" berths. Probably its most historic flight occurred in October 1950 when the VC-118 carried President Truman to Wake Island for discussions with Gen. Douglas MacArthur concerning the Korean War.

The Fairchild C-119 Flying Boxcar was designed to carry cargo, people, litter patients, and mechanized equipment. The C-119J on display has a unique history. It was especially modified for the midair retrieval of space capsules returning from orbit. On August 19, 1960, it made the world's first midair recovery of a space capsule when it snagged the Discoverer XIV parachute at 8,000 feet about 360 miles southwest of Honolulu.

Flying Boxcars saw much action during the Korean conflict as transports and returned to combat during the Vietnam conflict as AC-119 gunships in a ground support role. *U.S. Air Force photo.*

No, it's not W. C. Fields; it's the nose of a Douglas C-133A Cargo Master. Turboprop C-133s fulfilled the Air Force requirements for large capacity cargo aircraft throughout the 1960s. With its rear- and side-loading doors it was capable of handling a wide variety of military cargo. Most significant was its ability to transport ballistic missiles cheaper and faster than by trailer over highways.

The C-133A on display set a world record for propeller-driven aircraft on December 16, 1958, when it carried a payload of 117,900 pounds.

At the time of the 1954 French defeat in Southeast Asia, representatives of the major powers and of the Indochinese people met in Geneva and divided Indo-China into North Vietnam and South Vietnam, creating Laos and Cambodia at the same time. This section of the museum presents a part of the unsuccessful American attempt to keep South Vietnam free. The bulk of the U.S. Air Force combat support was flown from bases in Thailand, as indicated on the map.

This exhibit commemorates Airman First Class William H. Pitsenbarger who was posthumously awarded the nation's highest honor during a ceremony on December 8, 2000 at the museum. The pararescueman was awarded the Medal of Honor for treating and protecting scores of wounded infantrymen in a rain forest near Saigon in 1966 while under intense enemy fire and being mortally wounded himself. In the audience for the ceremony were battle survivors and hundreds of pararescue airmen.

Aircraft used in Vietnam ranged from the large B-52 bomber (*left*), which operated from bases in Thailand and Guam, to the small 0-1 and 0-2 forward aircontroller aircraft that flew in Vietnam. The camouflage-painted projectile was dropped along jungle roads to monitor enemy truck and troop movements.

Visitors in the Modern Flight Hangar may inspect the interior of this Douglas C-124C, one of 448 Globemaster IIs built in 1949-55. Visitors in the main museum complex also are able to walk through a B-29 bomber from WW II and sit in the cockpits of F-4 and F-16 fighters of the jet age. *U.S. Air Force photo by Harry Elliott.*

An Army utility aircraft is unloaded through the "clamshell" doors of a Globemaster in 1965 at Bien Hoa Air Base in Vietnam. "Old Shakey", as it was affectionately known, also could carry tanks, field guns, trucks, bulldozers, helicopters, or two hundred fully-equipped soldiers. The C-124 fleet was retired in mid-1974. *U.S. Air Force photo.*

DeHavilland Aircraft of Canada built the C-7A Caribou as a short takeoff and landing (STOL) utility transport, first flown in 1958. In Vietnam, its STOL capability made it particularly suitable for delivering troops, supplies, and equipment to isolated outposts. It could carry more than three tons. This combat veteran later served with the Air Force Reserve and was flown to the museum in May 1983. *U.S. Air Force photo.*

Deliveries of the Northrop YC-125B Raider began in 1950, but the Air Force soon found they were underpowered for use on rough, short airfields—their intended use. So they were employed mostly for ground maintenance training at Shepard AFB, Texas, and declared surplus in 1955. This Raider is painted to represent the YC-125 based at Wright-Patterson AFB in 1950 and used for cold weather testing. It went on display in 1995.

Aircraft designers modified the Boeing B-29's belly and produced a transport they designated the C-97. The first of the seventy-four Stratofreighters flew in November 1944. A tanker version was introduced in 1950 with 816 eventually being built. This KC-97L was christened *Zeppel inheim* in 1973 by a mayor of the German town, honoring its use by the Ohio National Guard during the aerial refueling of NATO forces in Europe.

The first Lockheed C-130 Hercules flew in August 1954. They were still being constructed in the late 1990s and used around the world by many nations. With the U.S. Air Force they have been equipped for midair space capsule recovery, ambulance service, drone launching, midair refueling of helicopters, reconnaissance, and radar weather mapping, as well as transports and gunships. This A model also was a test bed for numerous projects of the Air Force Systems Command.

Five Chance Vought XC-142s were built and only this one remains. The first of these unusual aircraft, which can fly straight up and down as well as forward, flew conventionally in late 1964. Vertical flight was achieved in January 1965, when the wings and engines were pointed skyward while the fuselage remained horizontal. The wings and engines were then tilted forward to attain horizontal flight. The four turbo-prop engines were linked together in such a way that one operating engine would turn all four propellers as well as the tail rotor. *U.S. Air Force photo.*

Four adjustable exhaust nozzles beneath the wing roots of the Hawker Siddeley XV-6A Kestrel could be rotated to provide thrust for vertical, backward, or hovering flight, as well as conventional forward movement. The United States received six of these aircraft for testing. When this Kestrel was flown to the museum from Edwards Air Force Base, California, in 1970 it became the first aircraft to be airlifted by the giant C-5A.

The massive radomes above and below the fuselage of Lockheed's EC-121 Constellation carry six tons of electronic gear. On October 24, 1967, over the Gulf of Tonkin, this unarmed plane guided a United States fighter by radar into position to destroy an enemy MiG-21. This was the first time a weapons controller aboard an airborne radar aircraft had ever directed a successful attack on an enemy plane. The EC-121D on display was nicknamed *Triple Nickel* because of its serial number.

Constellations over Southeast Asia also directed fighter-bombers to their aerial refueling tankers and guided rescue planes to downed pilots. EC-121s, such as this one shown over Thailand, evolved from the Lockheed commercial transport and from the USAF version that flew radar patrol off the United States coasts. *U.S. Air Force photo*.

Visitors have this view of President Eisenhower's VC-121E (*background*) and the Lockheed VC-140B Jet Star that served Presidents Nixon, Ford, Carter, and Reagan a number of times. President Eisenhower's wife christened this VC-121E the *Columbine III* in honor of the official flower of Colorado, her adopted home state. It was retired to the museum in 1966 while the Jet Star was flown to the museum in 1987. Only when the President is aboard an airplane, that particular plane flies under the radio call sign of "Air Force One."

A very special welcome ceremony was held at Wright Field on May 20, 1998, when this Boeing VC-137C was delivered to the Air Force Museum. It was made famous as "Air Force One" by President Kennedy, whose first flight aboard it was on November 10, 1962, with his wife to attend the funeral of former first lady Eleanor Roosevelt. One year later, on November 22, the four-engine jet flew President Kennedy to Dallas, Texas, where he was assassinated. On the flight returning his body, Vice President Johnson was sworn in as the 36th President. This aircraft also is known as "SAM 26000" for Special Air Mission and its tail number. *Photo by Philip Handeman.*

President Eisenhower was flown on this Aero Commander U-4B in 1956-60. Later it was used at the Air Force Academy for cadet parachute training and then by the Nebraska Civil Air Patrol. The museum acquired it in 1969 from a private source.

A B-29 fuselage is available for visitors to walk through and thus view crew work areas and bomb casings. This fuselage is painted in the markings of the *Command Decision* whose crew shot down five MiG-15s during the Korean Conflict.

This Convair B-36J made the last flight ever by a Peacemaker and was the first aircraft placed inside the museum. Because of its size, the B-36 had to be moved into the Air Power Gallery before construction was completed. It rested along the center line of the large structure while its 230-foot wing stretched its tips to within a few feet of the arched walls. The B-36 intercontinental bomber was on duty as a deterrent to aggression from June 1948 until it was replaced by the B-52 in 1959.

This type of thermonuclear bomb was carried by B-36 bombers from 1954 to 1957. The Mark 17 weapon is nearly 25 feet long and weighed 41,400 pounds. A B-36 is shown in flight at right.

U.S. Air Force

When the B-36 was designed during World War II, landing gears with huge single wheels were envisioned. As the design was perfected, the single wheels were replaced by four smaller wheels which better distributed the weight of the aircraft on the runway. Loaded, the Peacemaker weighed about 410,000 pounds.

Protection for the B-36 bomber was the mission of the little McDonnell XF-85 Goblin, seen next to its mother ship. It was to have been carried inside the B-36 and launched from a trapeze to fight attackers with its four .50-caliber machine guns. Afterward it would fly under the bomber, re-attach itself to the trapeze, fold its wings, and be lifted back into the bomb bay. The project was cancelled in favor of midair refueling of normal-sized fighters by tankers. The Goblin first flew in August 1948 and was transferred to the museum two years later.

The Boeing B-47 Stratojet is the world's first swept-wing bomber. It also is the first airplane built solely for the delivery of nuclear weapons. In December 1953 it set a new transatlantic speed record averaging 650.5 miles per hour. The B-47E on display was the first Air Force aircraft to incorporate a "fly-by-wire" primary flight control system in which the pilot's command controls were transmitted to the control surfaces by electrical wires rather than by cables and mechanical linkages.

Maj. Colin A. Clarke flew this LTV A-7D Corsair II on a nine-hour rescue support mission in Southeast Asia on November 18, 1972, for which he earned the Air Force Cross—one level down from the Medal of Honor.

Capt. Paul Johnson flew this Fairchild Republic A-10A Thunderbolt II on an eight-hour rescue support mission during Operation Desert Storm on January 21, 1991, for which he earned the Air Force Cross. The A-10 was built specifically to attack tanks and other armored vehicles.

Lucky Lady II was the name of the first airplane to fly nonstop around the world. It was a Boeing B-50 Superfortress and the 23,452 mile trip required four aerial refuelings. The B-50 was the last propeller-driven bomber to be delivered to the Air Force. Originally called the B-29D, it had more powerful engines, a taller rudder, which could be lowered to the side for maintenance, a new wing structure, a new undercarriage, and hydraulic nosewheel steering. A WB-50D is on exhibition.

Used primarily as the avionics test bed for the B-1B, this Rockwell International B-1A Lancer was flown to the museum in December 1986 soon after the B model became operational. It is painted now as it was during the early phases of the B-1A flight test program that started in 1979. Lancers are expected to be available for high-speed, low altitude missions until about the year 2037. To the right of the B-1A is a Convair C-131D Samaritan.

Three Boeing B-52B Stratofortresses in January 1957 set the first non-stop jet-flight record around the world in 45 hours, 19 minutes. In December 1972, the B-52D on display (*center, background*) flew four long-range bombing missions over North Vietnam to help free American prisoners of war during the massive demonstration of U.S. resolve by President Nixon. Nearly twenty years later, in January 1991, B-52 veterans of Vietnam hastened the Iraqi defeat during Operation Desert Storm. At the end of the decade they were being used again to support the NATO air campaign over Yugoslavia. Early in the new century, B-52s were bombing Terrorists' strongholds in Afganistan. Also shown with the B-52D here is a General Dynamics F-16A painted as a "Thunderbird" of the USAF Aerial Demonstration Team. This Fighting Falcon flew with the team for ten years to 1992.

Aerial refueling by the Boeing KC-135 Stratotanker has given the B-52 its around-the-world capability. *U.S. Air Force photo.*

Great Britain developed this airplane and the United States borrowed its design, calling it the Martin B-57 Canberra. It served with the Tactical Air Command starting in 1954 and found a new lease on life in Vietnam as a low-level bomber and fighter.

U.S. Air Force photo.

The B-66 was the last tactical bomber built for the Air Force, and only the B-66B was designed exclusively as a bomber. Others served as tactical reconnaissance aircraft, while the final version, the WB-66D, was designed for electronic weather "recon." The Douglas RB-66B Destroyer on display flew combat missions in Southeast Asia in an electronic countermeasures role.

U.S. Air Force photo.

The Air Force's first supersonic operational bomber was the delta-wing Convair B-58 Hustler. The plane's wasp-waist fuselage left no room for bombs so a droppable two-component pod was carried beneath it. The pod contained extra fuel and a nuclear weapon, reconnaissance equipment, or other specialized gear. The B-58A at the museum set three separate speed records and earned the crew the Bendix and Mackay trophies for 1962 during a flight from Los Angeles to New York.

U.S. Air Force photo.

Prior to 1988 as visitors arrived at the museum area, one of the first individual planes they saw was the North American XB-70 Valkyrie thrusting out toward them looking for all the world like an oversized platypus. The XB-70 was designed to make use of a little-known phenomenon called "compression lift." This lift is achieved when the shock wave generated by the shape of an aircraft flying at supersonic speeds actually supports part of the airplane's weight. Built largely of stainless-steel honeycomb sandwich panels and titanium, the Valkyrie could drop its wing tips as much as 65 degrees at supersonic speeds to improve its stability. This XB-70 is the last of the two that operated from 1964 to 1969. Most aircraft in the background and the XB-70 were moved in the late 1980s to various indoor display areas.

U.S. Air Force photo.

Northrop's Gee Whizz Deceleration Sled accelerated to 200 miles per hour along a 2,000-foot track, then stopped in 45 feet. The sudden stop developed deceleration forces such as those encountered in airplane crashes and the opening of parachutes. The tests were begun in 1947 for the Aeromedical Laboratory at Wright Field.

Bell designed and built six airplanes in the X-1 series, the last of which is the X-1B displayed at the museum. This flying laboratory was equipped with 1,-000 pounds of complex instruments to answer questions about the then strange new world of the sound barrier. Capt. Charles E. Yeager broke the sound barrier in an earlier model in a straight and level flight on October 14, 1947. Six years later he flew the X-1B to a record 1,650 miles per hour.

Douglas made only one X-3 Stiletto. It was designed to test aircraft features at supersonic speeds and high altitudes. Data gained from the X-3 program of the mid-1950s was of great value in the development of the F-104, X-15, and other high-performance aircraft. Two of the minihelicopters displayed behind the X-3 were self-propelled. The Nazi observation helicopter in the center was towed behind a submarine.

Mercury, Gemini, and Apollo space projects received significant contributions from the X-15 flight program. Several X-15 pilots earned astronaut ratings by flying above fifty miles and into the lower edges of space. The three North American X-15s which were built made 199 flights between 1959 and 1968. The X-15A-2 displayed set an unofficial world speed record of 4,520 miles per hour in August 1966.

From the museum floor, the forward-sweep wing and forward mounted elevator (canard) are obvious features of the Grumman X-29A. It was built in the early 1980s to explore state-of-the-art technologies for future aircraft design. It was presented in 1995 to the museum by the Wright-Patterson Flight Dynamics Laboratory.

This Ryan X-13 Vertijet (*right*) demonstrated on April 11, 1957, at Edwards AFB, California, that a jet aircraft could takeoff vertically, transition to horizontal flight, and return to vertical flight for landing—essentially on its tail. It was transferred to the museum two years later. In the background, above the B-58 tail, are a Cessna T-37B "Tweety Bird" and a propeller-driven Cessna T-41A Mescalero.

★ 7 | *Missiles and Space*

As visitors continue their chronological walk through aviation history in the vastly-expanding U.S. Air Force Museum, they might be able to conclude it with visits to the vertical Hall of Missiles and the new Space Gallery. In the enlarged Space Gallery, visitors again will be able to see gondolas which, suspended from huge balloons, ascended to new heights; an actual capsule that carried Americans into space; samples of food eaten by astronauts; exotic equipment recovered from early space tests; full-scale satellites hanging dramatically overhead; and, for the first time, "new" space items from the museum's extensive storage facilities. Temporarily, some items are to be displayed in the Eugene W. Kettering Building.

Visitors also will be able to enjoy a ten-minute video on the Air Force and NASA as partners in space, complete with captions for the hearing impaired. And in an especially protected display case, visitors may inspect an actual moon rock only inches from their eyes.

For the most part of three decades, in front of the museum were displayed the giant space missiles that have helped protect the country and support the nation's peaceful exploration of space. These were removed in early 1997 and placed in storage for later restoration and eventual exhibition inside the missile silo that is being built adjacent to the Kettering Building. Several smaller missiles and mobile vehicles to launch other missiles temporarily were to remain displayed outdoors.

When the Air Force launched its first satellite in 1958, it took the lead in developing space systems to improve national security. As the Department of Defense's principal space agency, the Air Force supports the Army, Navy, and NASA. About two-thirds of the payloads placed in earth, moon, or sun orbit by the United States in the early decades were launched by the Air Force.

While developing space technology for the future, the Air Force and the

154

nation's space programs have developed technology for today. Weather and communications satellites have more than proven their worth. And satellites have become nearly indispensable for surveillance, warning, mapping, education, and ecology management. In fact, thousands of new products and techniques for military and civilian use have resulted from space and missile development. A number of these new products were displayed in the museum's original Space Gallery, as they will be in its successor. From there, visitors could return to other areas of the museum or, if they had not allowed enough time, make plans for a return visit.

A family examines a moon rock brought to Earth in 1972. In the Kitty Hawk panel are fragments of the 1903 Wright Flyer that Neil Armstrong took to the moon in 1969. Overhead, as shown in this original Space Gallery scene, is a Cessna 0-2 with a service ceiling of 19,300 feet.

Visions of space travel can be conjured up by visitors in this section of the Space Gallery. In fact, the Apollo 15 Falcon command module (*right*) took three Air Force astronauts to the moon in 1971 on a twelve-day mission. Maj. Alfred M. Worden, Jr. remained in the capsule in moon orbit while Col. David R. Scott and Lt. Col. James B. Irwin explored the moon. The development of space suits is depicted on the left.

This Manhigh II balloon gondola was used in a project established in December 1955 to obtain scientific data on the behavior of man in an environment above 99 percent of the earth's atmosphere, and to investigate cosmic rays and their effects upon man. Three balloon flights to the edge of space were made during the program.

Project Stargazer was established in January 1959 for high-altitude astronomical investigation from above ninety-five percent of the earth's atmosphere. This permitted undistorted visual and photographic observations of the stars and planets. On December 13/14, 1962, Capt. Joseph Kittinger and Mr. William White made a balloon flight to an altitude of 82,200 feet over New Mexico in a Stargazer gondola. The flight also provided valuable information concerning the development of pressure suits and associated life support systems.

Captain Joseph Kittinger also ascended alone to an altitude of 102,800 feet in an Excelsior open gondola in August 1962. He set two balloon ascension records on the way up and two parachute records on the way down. He later served as a fighter-bomber pilot in Southeast Asia, was captured by the North Vietnamese, and was freed during "Operation Homecoming" in 1973.

CAPT. KITTINGER IN GONDOLA DURING LAUNCH FROM HOLLOMAN AFB N. MEX.

PHOTO OF CAPT. KITTINGER BY AUTOMATIC CAMERA AS HE PREPARED TO JUMP FROM 102,800 FEET

FALLING PRIOR TO DEPLOYMENT OF STABILIZING CHUTE

LANDING ON THE RUGGED NEW MEXICO DESERT

Two Philippine monkeys and two white mice were launched to an altitude of 36 miles on May 22, 1962, to help man prepare for his journeys into space. They traveled skyward at 2,000 miles per hour aboard an Aerojet Aerobee rocket and returned safely to earth in its nose section dangling beneath a parachute.

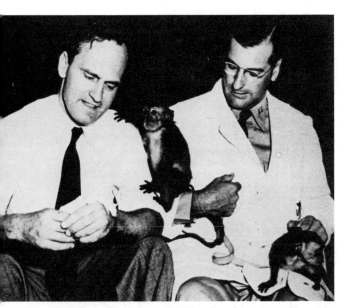

Monkeys Patricia and Mike thus became the first primates to reach so high an altitude. The white mice, Mildred and Albert, romp on the hands of the gentleman in the white shirt. Their flight provided invaluable data for the survival of man during rocket launch as well as man's existence in the weightless environment of space.

A monkey doll illustrates in two views how Patricia traveled while aboard the Aerobee. Mike rode in a lower compartment—on his back.

This Lockheed Research Satellite Engineering Development Model was designed to carry several types of scientific payloads in order to make it more versatile in investigating space. Maximum weight at launch was 350 pounds. The first launch of this type satellite was in 1963 from an Agena space vehicle already orbiting the earth. During hundreds of orbits, the satellite radioed back 100,000,000 measurements of scientific data.

Closeup of Northrop OV-2-5 space satellite.

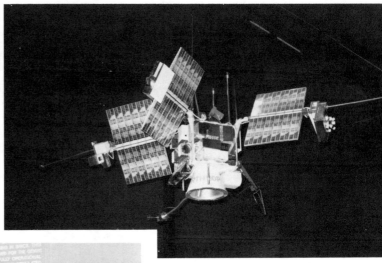

Space Shuttle operations stimulated the need for such equipment as this extravehicular activity "E-V-A" unit to repair items in space. The backpack originally was designed for the Gemini program.

This array of missiles faces the IMAX Theater portion of the Air Force Museum. The three units are (*left to right*) the Boeing CIM-10A Bomarc, Martin TM-61A Matador, and the Martin CGM-13B Mace. While the Bomarc was designed to intercept and destroy enemy aircraft, the Matador and Martin were designed to destroy ground targets.

Signs readily identify missiles, aircraft, and other vehicles both indoors and out—thus facilitating self-guided tours at one's own pace.

MARTIN CGM-13B "MACE"

Like its predecessor, the Matador, the Mace was a tactical surface-launched missile designed to destroy ground targets. It was first designated as the TM-76 and later as the MGM-13. It was launched from a mobile trailer or from a bomb-proof shelter by a solid-fuel rocket booster which dropped away after launch; a J33 jet engine then powered the missile to the target. Development of the Mace began in 1954 and the first test firing occurred in 1956.

The Mace was developed in two versions, the "A" and the "B." The "A" employed a terrain-matching radar guidance system known as "ATRAN" (automatic terrain recognition and navigation) in which the return from a radar scanning antenna was matched with a series of radar terrain "maps" carried on board the missile and which corrected the missile flight path if it deviated from the film "map." The "B" used a jam-proof inertial guidance system; it had a range twice that of the "A."

Mace "A" missiles were first deployed to USAF forces in Europe in the spring of 1959. These remained in service until the mid-1960s, when some were used as target drones because their size and performance characteristics resembled those of a manned aircraft. Development of the "B" missiles began in 1964 and these remained operational in Europe and the Pacific until the early 1970s. The Mace "B" on display was based on Okinawa prior to its delivery to the Museum in 1971.

SPECIFICATIONS

Span: . . . . 22 ft. 11 in.
Length: . . . . 44 ft. 6 in.
Height: . . . . 9 ft. 7 in.
Weight: 18,000 lbs. at launch
Armament: . . . Conventional or nuclear warhead
Engine: . . Allison J33 of 5,200 lbs. thrust; Thiokol solid-propellant booster rocket of 100,000 lbs. thrust
Cost: $452,000

PERFORMANCE

Max. speed: . . . . . 650 mph/ 565 knots in level flight; supersonic in final dive
Range: . . 1,400 statute miles/ 1,217 nautical miles
Operating altitude: From under 1,000 ft. to 40,000 ft.

As a surface-to-surface missile, the Martin CGM-13 Mace was in operation from 1959 until the early 1970s. It used a guidance system that permitted a low-level attack by matching a radar return with radar terrain maps.

MARTIN TM-61A "MATADOR"

The Matador was a surface tactical missile designed to carry either a conventional or a nuclear warhead. Originally designated as the B-61, the USAF's first "pilotless bomber," it was similar in concept to the WWII German V-1 "buzz bomb." The Matador was launched by a booster rocket from a mobile 40-foot trailer and was controlled electronically from the ground during flight. Immediately after launch, the booster rocket fell away and the missile continued on course to target powered by its jet engine.

Development of the Matador began in August 1945 and the XB-61 was first launched on January 19, 1949. Operational TM-61s which later followed were the first tactical guided missiles in the USAF inventory. The first Pilotless Bomber Squadron (Light) was organized in October 1951 for test and training purposes and in March 1954 the first Matador unit was deployed overseas to bolster NATO forces in West Germany. TM-61 units were also sent to Korea and Taiwan.

Martin delivered the 1,000th Matador in mid-1957, but in 1959 a phase-out of the Matador began in favor of a more advanced version, the Martin "Mace."

SPECIFICATIONS

Span . . . . 27 ft. 11 in.
Length . . . . 39 ft. 8 in.
Height . . . . 9 ft. 8 in.
Weight: 13,593 lbs. (at launch)
Armament . .Conventional or nuclear warhead
Engines: Allison J33 of 4,600 lbs. thrust; Aerojet solid-propellant booster rocket of 57,000 lbs. thrust

PERFORMANCE

Maximum speed . . . 600 mph (level flight; supersonic during final dive)
Range . . . . 690 miles
Service ceiling . . . . 44,000 ft.

Cost $132,000

Testing of the Boeing CIM-10 Bomarc began in 1952, became operational in 1960, and was phased out by 1972. It was a surface-launched pilotless interceptor, which was designed to destroy enemy aircraft. The "A" model shown here had a range of 260 miles. Improved "B" models were stationed in the United States and Canada.

These silent sentinels once stood as if at attention in front of the museum. Since early 1997 they have been in storage. The are (*left to right*): Boeing LGM-30G Minuteman III, Boeing LGM-30A Minuteman I, Chrysler PGM-13 Jupiter, Martin HGM-25A Titan I, Douglas PGM-16 Thor, and the Convair HGM-16F Atlas.

This is an artist's concept of the interior of the new home of the vertical missiles that once were displayed outdoors. Construction on the above-ground silo began in October 2002. *U.S. Air Force photo.*

Research data for the development of an intercontinental winged missile was gathered by the North American X-10. Although thirteen of these supersonic vehicles were built and flown in the mid-1950s, only two are known to exist. *U.S. Air Force photo.*

This type Aerojet rocket engine powers the second stage of the Minuteman 3 ICBM (intercontinental ballistic missile). After a Minuteman is launched by its larger first-stage engine, the first-stage section is jettisoned when its fuel is depleted, and the second-stage engine takes over to boost the missile higher and faster. It, likewise, is jettisoned, and the third-stage engine then takes over to propel the nuclear warhead onward toward its preselected enemy target.

The Chrysler PGM-19 Jupiter intermediate-range ballistic missile (IRBM) was developed by the Army's Ballistic Missile Agency under Dr. Wernher von Braun. When it became operational in 1959, it was placed under Air Force control. Originally designed as the SM-78, the Jupiter was a single-stage, liquid-propellant missile using an all-inertial guidance system. Jupiter squadrons of fifteen missiles each were deployed at NATO launch sites in Italy and Turkey in 1961. As more advanced missiles were developed, the Jupiter became outdated and in 1963 it was withdrawn from military use. Some Jupiters were used as first-stage boosters to launch space satellites.

Liquid propellants for the Martin HGM-25A Titan I's Aerojet rocket engines were kerosene fuel and liquid oxygen. The Titan I was the first Air Force ICBM to be placed in hardened underground silos for protection against enemy attack. However, they had to be lifted from their silos to the surface by elevator prior to launching. By 1965 Titan Is were being phased out in favor of Titan IIs, which offered greater range and payload and which were launched from within their silos.

"One of the most versatile USAF space vehicles" is how this Lockheed Agena A has been described. Such upper stage vehicles were used on boosters including the Thor, Atlas, and Titan IIIB. The two panels on the right (*above*) tell of Dr. Robert H. Goddard's early rocket experiments in New Mexico and of the Germans building on his successes for the V-2 rockets that they launched against England in WW II.

The cone-shaped exhibit officially is know as the ASV-3 ASSET Lifting Body. It is the only one of six to survive a 1963-65 program designed to develop a reusable, maneuverable, re-entry vehicle capable of being flown from earth orbit to a precise landing point on earth. McDonnell Aircraft built the ASSETs, which were launched on Thor boosters to an altitude of about 195,000 feet.

In order to thwart Soviet planners during the Cold War, the Air Force in 1989-91 tested this unusual tractor-trailer rig. The Hard Mobile Launcher could travel at 55 mph on the highway and, safely away from any incoming Soviet missiles, could launch in retaliation a "Midgetman" Small Intercontinental Ballistic Missile. Behind the rig is a Peacekeeper Rail Garrison Car from which a Minuteman ICBM could be fired. Both projects were cancelled in 1991 after Cold War tensions eased. They may be viewed at the far end of the Outdoor Air Park at the museum.

The Titan II gyroscope, which is preprogrammed by a computer installed in the missile, steers with such accuracy that it will pinpoint a target 10,000 miles from the launch site. The gyroscope is gold-plated because gold is the best-known material for withstanding corrosion and temperature changes encountered in space.

The Northrop OV-2-5 space satellite mock-up is shown above a F-94A which had a service ceiling of 42,750 feet. The satellite orbited at 22,200 miles above the earth. It was designed for a year of solar, magnetic, and cosmic ray research in space. On September 28, 1968, it was boosted from the ground by a Titan III launch vehicle.

A Minuteman is launched at sunset from Vandenberg Air Force Base, California. *U.S. Air Force photo.*

The Discoverer XIV capsule on display was the first item to be ejected by a satellite orbiting in space and to be recovered in midair. It was launched August 18, 1960, by a Thor booster and propelled into orbit by an Agena. Upon descent and reentering the earth's atmosphere it released a parachute which was snagged by a C-119 recovery airplane. Discoverer XIV thus completed a 27-hour, 450,000-mile journey through space.

The Convair HGM-16F Atlas, the Free World's first ICBM, was designed to be launched at an enemy target at least 5,000 miles away. Lt. Col. John Glenn was put into orbit on February 20, 1962, by an Atlas missile. It also has been used to launch a variety of unmanned spacecraft including Ranger, Surveyor, and Mariner.

As seen through a glass doorway, two "officers" work inside a Boeing Minuteman II Mission Procedures Trainer. Thirteen weeks in such a trailer were required at Vandenberg AFB, California, then more at the launch site, and monthly training and evaluation afterwards. The YRF-84F cockpit and Hangar 9 doors are reflected in the glass.

This is an Atlas ICBM being launched from Vandenberg Air Force Base, California, in February 1962.

U.S. Air Force photo.

Artist: Russ Turner

General Bernard A. Schriever spearheaded the development of the Air Force's early accomplishments in space. His portrait is part of the Air Force Art Collection at the museum.

Some of the attractions in the original Space Gallery were (*left to right*) the Apollo 15 command module, Gemini and Mercury spacecraft, tail of the X-24A, space foods display, and a space theater with sound and captions. *U.S. Air Force photo.*

This type of space sled was once considered for travel near orbiting spacecraft. Eventually, a back-pack design was chosen for such extravehicular activity.

One man rode into space in this type of McDonnell Mercury spacecraft (*right*) during the six Project Mercury flights from 1961 to 1963. Although the McDonnell Gemini spacecraft (*left*) was six inches shorter, it was designed to carry two men into space. Ten manned Gemini missions were flown in 1965 and 1966 as the bridge to landing on the moon with the Apollo module. The Mercury on display was used to provide parts to support the flight of L. Gordon Cooper, Jr. on May 15-16, 1963 while the Gemini on display was used for thermal-qualification testing. *U.S. Air Force photo.*

Two lifting bodies, the Martin X-24A (not shown) and the X-24B (shown here), helped pave the way for the design and construction of the Space Shuttle, which was being tested in 1977. (A lifting body is an aircraft that derives lift from the shape of its fuselage rather than from wings. It is thus better able to cope with the heating associated with high speeds.) The two lifting bodies made sixty-four powered flights from the wing of a B-52 between March 1970 and November 1975. The program was managed by the Air Force Flight Dynamics Laboratory at Wright Field.

One Thiokol XLR-11 rocket engine of 8,000-pounds thrust and two Bell LLRV optional landing rockets of 400-pounds thrust each powered the X-24B. It had a maximum speed of 1,163 miles per hour.

The X-24A was modified into the X-24B in 1972 when the testing of the former was completed. The X-24A on display was built as the jet-powered SV-5J for flight training, but never flown. It was converted to the X-24A configuration for display purposes. *U.S. Air Force photo.*

★ *8* Special Exhibits

Unique exhibits and some of the U.S. Air Force Museum's smaller displays are on occasion presented to the public in the two exhibit halls that flank Carney Auditorium in the center of the front building. Some are special one-time exhibits, some are smaller ones that remain in the same general area, and some are later moved near to the aircraft or historical era to which they are related. At times the personalities associated with the exhibit attend the public unveiling ceremony.

Two spectacular walk-through exhibits were featured as the twentieth century drew to a close: the fiftieth anniversaries of the independent Air Force in 1997-98 and of the Berlin Airlift in 1998-99. Those were followed in 2000-02 with the fiftieth anniversary of the Korean War, when the Air Force came of age. Each included many life-size dioramas, video presentations, theatrical lighting, and sound effects. Two decades before, a large number of black airmen who had flown in World War II participated in the dedication of a free-standing display that recognized their contributions in all conflicts. Much of its contents were subsequently incorporated into other wall exhibits as they were expanded.

As noted in Chapter 1, this museum core area has a variety of displays on the first and second floors. They include paintings from the Air Force Art Collection and special shows by aviation artists for both art lovers and flying history buffs; the Eugene W. Kettering collection of model airplanes; drawings by Woodi Ishmael of Medal of Honor recipients; enlisted award winners; history of the Bendix Trophy; recognition of contributions to aviation by cartoonist Milton Caniff; and special-theme exhibits. The Student Aviation Art Show is hung as well each Spring in the museum.

Exhibits, of course, are located throughout the museum, in the Eighth Air Force control tower, the two adjacent Nissen Huts, and—from a different perspective—in the Memorial Park. The openings of new exhibits and the presentations of newly acquired aircraft are announced in advance through the news media so that interested persons may attend.

The nation's highest military award, the Medal of Honor, has been presented to at least sixty-two airmen. This includes Special Congressional Medal Awards to Capt. Charles A. Lindbergh, Brig. Gen. William "Billy" Mitchell, Brig. Gen. Charles E. Yeager, and Lt. Gen. Ira C. Eaker. General Eaker received his award in 1979 on the seventy-sixth anniversary of powered flight. Normally, the Medal of Honor is awarded for "conspicuous gallantry and intrepidity at the risk of life above and beyond the call of duty." It was awarded to four Air Force men in World War I, thirty-eight in World War II, four in the Korean conflict, and twelve for combat in Southeast Asia. The distinctive Air Force design (*right*) was created in 1965. Maj. Bernard F. Fisher was the first to receive it on January 19, 1967.

Many of the museum's smaller displays are introduced to the public in the gallery that houses the Medal of Honor exhibit. Here in words and pictures is the story of "Aerial Photography and the Cuban Missile Crisis" of 1962. Supported by photographic evidence and a strong military force, President Kennedy persuaded the Soviet Union to withdraw its offensive missiles from Cuba.

This Photographic Heritage on Film Gallery featured rotating exhibits on a variety of themes pertaining to aviation history. Nearly all of the photographs were enlargements from prints in the museum's vast files. The gallery also offered visitors an opportunity to relax while viewing aviation personalities in their history-making roles.

The Civil Air Patrol was established on December 1, 1941, by decree of the Director of Civilian Defense Fiorella H. La Guardia. The CAP is an official auxiliary of the United States Air Force. As such, it carries out an education program for youth and assists with the search and rescue of downed airplanes, both military and civilian.

For three decades, the annual race for the Bendix Trophy produced big headlines across the country. Maj. James H. "Jimmy" Doolittle in 1931 won the first race, averaging 223 mph from Los Angeles to Cleveland, flying a Laird Solution. An Air Force B-58 crew in 1962 won the last race, averaging 1214.71 mph from Los Angeles to New York. Their Convair Hustler was retired to the museum seven years later.

Over the decades, various Air Force and Air National Guard aerial teams have demonstrated the performance capabilities of operational aircraft and the skill of military pilots. This exhibit presents the history of the Thunderbirds and other aerial demonstration teams back to the Air Corps' Three Musketeers of 1928.

Col. Bernt Balchen was America's greatest Arctic expert of modern times. The Norwegian native was the chief pilot in 1928-31 for Admiral Byrd's Antarctic Expedition and on November 29, 1929, Balchen piloted the first aircraft to fly over the South Pole. After serving in World War II, he was recalled to duty in 1948 and in the following year made the first nonstop flight over the Polar region from Alaska to Europe. Lowell Thomas dubbed him the "Last of the Vikings."

The history, traditions, and uniforms of the Air Force Academy are highlighted in this special exhibit. Located at Colorado Springs, the four-year institution produces new second lieutenants with college degrees.

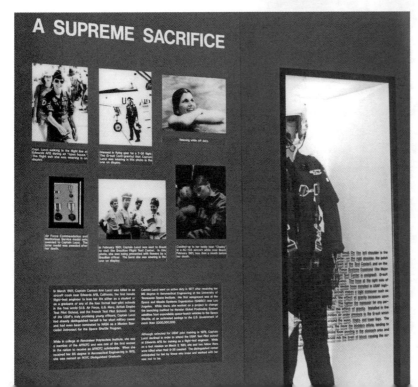

This display remembers Capt. Carmen Ann Lucci who was killed in 1961 in an aircraft crash near Edwards AFB, California. She thus became the first female flight-test engineer to lose her life either as a student or as a graduate of any of the four formal test-pilot schools in the Free World.

This special exhibit opened at the museum in March 2002 to salute Bob Hope for fifty-plus years of service to the U.S. military and the nation. Among those at the opening were his daughter Linda and his son Kelly, special guest Phyliss Diller, and veterans who had been entertained by Bob Hope in World War II, Korea, Vietnam, and the Gulf, as well as over the decades within the United States and aboard ship.

"City Held Hostage: Berlin" was the theme of this spacious walk-through exhibit commemorating the fiftieth anniversary of the Berlin Airlift. It took visitors through oppression, rubble, and destruction while telling the story of the Allies' humanitarian flights to keep West Berlin alive and free from Communist control.

During the Berlin Airlift, officially "Operation Vittles," Lt. Gail S. Halvorsen started the custom of parachuting candy to children. Here one such youngster retrieves a parcel that had landed in a tree. A German businessman who toured the exhibit told his guide that as a child he had chased down such treats.

Museum visitors walking through the exhibit highlighting the fiftieth anniversary of the Korean War in 1950-53 entered along side of two pilots at the Tori Gate leading to the F-86 flight line at Kimpo Air Base. The massive exhibit could be viewed from floor level or from the parallel gallery above.

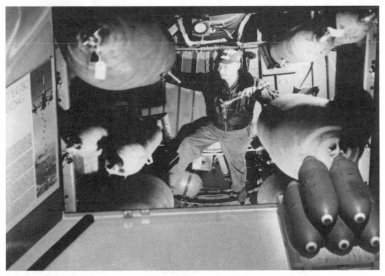

Giant murals were used throughout the exhibit to illustrate the many roles of the Air Force in the Korean War. This B-29 crewman could be preparing his "Superfortress" for close support of Army ground troops or to bomb strategic bridges.

Time to go home.

Capt. Mary Spivak was among the flight nurses who assisted C-54 crews in transporting 1,000 orphans from Seoul to safety on an island off the southern coast of Korea. The Chinese offensive of 1950 had threatened to overrun the youngsters' orphanage. For more on "Operation Kiddy Car," see page 123.

"Fabric of the Air Force" is the title of this huge quilt which commemorates the fiftieth anniversary of the Air Force. The 100 hand-made squares recognize past and current air bases. Thread used to join the squares was flown on the September 1997 mission of the space shuttle Altantis.

Two residents of East Berlin are shown escaping to the west over actual sections of the Berlin Wall in this special exhibit. They used the roof of a Communist-made car to help them reach freedom during the Cold War. These artifacts are slated for perma-nent display in the new Eugene W. Kettering Gallery.

The Air Force Art Collection contains over 8,000 paintings, which are displayed world-wide. Nationally, there are permanent dis-plays at the Smithsonian Institution, the Air Force Academy, and the Air Force Museum. The pictures, which are rotated from time to time, are in two categories: Historical, 1915 to 1953; and Contemporary, 1954 to the present.

The New Look. Artist: Bill Edwards

Warming Up—Guam to Vietnam was painted on a board from a box which had contained bomb fuses. Some of the original lettering was purposely permitted to show through the paint by the artist, Howard Koslow.

Many members of the Society of Illustrators donate paintings and drawings to the Air Force Art Collection. Here two artists are shown gathering ideas at Ent Air Force Base.

Artists at Ent. Artist: Dale B. Gallon

Artist: Maxine McCaffrey

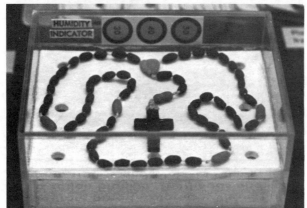

Many museum visitors are attracted to the Southeast Asia prisoner-of-war display of which this picture is a part. Several items were contributed by former POWs who received their medical and administrative processing during "Operation Homecoming" at Wright-Patterson Air Force Base.

This rosary is typical of those made in North Vietnamese POW camps. It was secretively made in 1971 or 1972 by Navy Comdr. Paul Schultz and given to Air Force Col. John P. Flynn, who smuggled it from North Vietnam during the POW release program.

A different art form, a bronze likeness of the legendary Icarus greets visitors in the main lobby. The statue was donated to the museum by alumni of the Air Force Institute of Technology at Wright Field. It remembers the men and women graduates who have given their lives in the service of our nation. According to Greek mythology, Icarus fell to his death when he flew too close to the sun, melting the wax on his artificial wings.

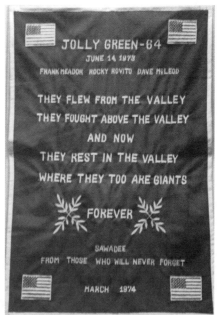

Maj. Emmett E. Hatch, Jr., an HH-3E pilot, wore this "Jolly Green" party suit at DaNang Air Base in South Vietnam when his helicopter rescue squadron would take a break from the realities of war. Those realities included honoring three comrades—Frank Meador, Rocky Rovito, and Dave McLeod—who died in 1973 when their helicopter crashed in a Cambodian lake. "Sawadee" is Thai for "goodbye."

"That others may live" is the motto of members of the Air Rescue Service. Their story, which includes help from the U.S. Navy, was told along two walls behind a collection of Air Force helicopters and amphibious aircraft.

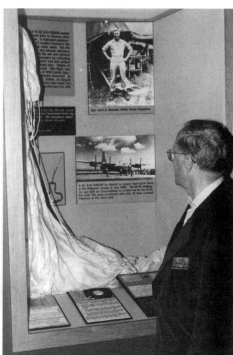

Museum volunteer Jack Munsell views the parachute and "D" ring that helped save his life when he bailed out of a B-32 on one of the last operational missions of WW II. He was rescued by the Navy.

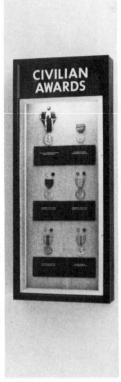

CIVILIAN AWARDS

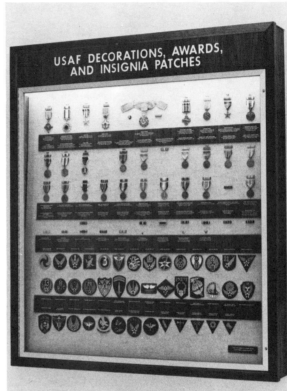

USAF DECORATIONS, AWARDS, AND INSIGNIA PATCHES

THIS TREE
IS A LIVING TRIBUTE TO
PRISONERS OF WAR AND
MISSING IN ACTION
IN SOUTHEAST ASIA
23 OCTOBER 1972
DONATED BY
VOLUNTEERS FOR POW/MIA

Recognition for bravery or for a "job well done" is given to civilians as well a active duty members of the Air Force. The lower half of the military displa shows various unit insignia patches. These display cases were among sever: throughout the museum that were temporarily removed for new exhibits.

EVOLUTION OF USAF AIRCRAFT INSIGNIA

Changes in the design, size, color, and location of national insignia on Ai Force aircraft have been the result of technical advances, mission changes and combat experience. During World War II, for example, the red circle within the white star was removed to avoid confusion with the rising sur insignia of the Japanese.

SOMALIA   SOUTH AFRICA   SPAIN

UNION OF SOVIET SOCIALIST REPUBLICS

UNITED ARAB REPUBLIC

UNITED KINGDOM

A few of the foreign wings displayed in a Wings of the World case are those from South Africa, Spain, Union of Soviet Socialist Republics, United Arab Republic, and the United Kingdom. A twin case not seen here features American wings and badges, past and present, including pilot's, navigator's, missileman's, flight surgeon's, nurse's, and parachutist's wings.

Humor has always been a part of the American experience, and so it is with Air Force life. Some of the lighter moments in the history of the nation's air arm are illustrated here.

A very important part of flying that the average person is not too concerned with is flying safety. This case displays several of the more humorous Blunder Trophies for safety violations. None of them is explained; however, the astute observer is able to conjure up several ideas.

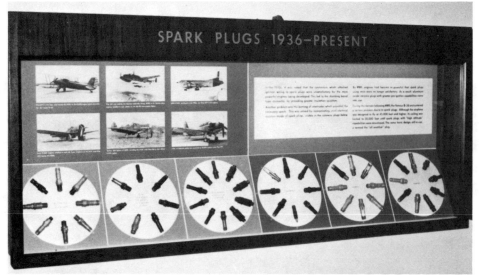

Automobiles and airplanes have a least one key, common element. Both depend on spark plugs. This display is one of three. The others are spark plugs to 1940 and jet and rocket engine igniters. Prior to World War I most of the plugs used in the United States were made in Europe. When the famous B-36 was constructed, it was designed to fly at 45,000 feet, but it was limited to 30,000 feet until a new type spark plug was developed. While a spark plug works throughout a flight, jet and rocket engine igniters work only to start their engines.

A collection of approximately 600 aircraft models was presented to the Air Force Museum Foundation for permanent display by the late Eugene W. Kettering. Started in mid-1920s, this collection reflects the significant advancements of civilian and military aircraft of many nations. It is a capsule history of aviation in model form, 1-76th the size of the full-scale aircraft. The models are all handmade of balsa wood and were once displayed in the study of Mr. Kettering's home.

Enlisted Heritage is the theme of this exhibit that portrays men and women in a variety of uniforms worn from 1907 to today. It has been retained from the much larger exhibit that celebrated the fiftieth anniversary of the independent Air Force.

The celebrated "High Flight" poem was written by Pilot Officer John Gillespie Magee, Jr. on the back of a letter (*above, bottom left*) he had written to his parents in August or September 1941. Magee was an American who was born to missionary parents in Shanghai, China, and was killed at the age of nineteen while flying with the Royal Canadian Air Force in December 1941 over England. His gravemarker there is shown top right. The poem begins "Oh, I have slipped the surely bonds of earth" and ends "put out my hand, and touched the face of God."

Many celebrities served in uniform during World War II. Those saluted on these panels are (*right to left*) President Ronald Reagan, James Stewart, Clark Gable, Gene Raymond, Jackie Coogan, and Joe Lewis.

The U.S. Air Force Song evolved from a 1938 contest sponsored by "Liberty" magazine. Of 757 scores submitted, a committee of Air Corps wives selected one written by Robert MacArthur Crawford, who introduced it at the 1939 Cleveland Air Races. The first page of his music was taken to the moon on July 20, 1971 aboard the Apollo 15 "Falcon" lunar module.

American prisoners of war and those listed as missing in action were honored by then Vice President Bush in 1988 when he dedicated this Missing Man monument. It was sponsored by the Red River Valley Fighter Pilots Association, a group named after a region that saw heavy action during the war in Southeast Asia.

Seventeen of the twenty-nine surviving Doolittle Tokyo Raiders were at the museum when this monument was dedicated on April 17, 1999. Names of all 80 participants are listed on the other side.

Individuals and organizations sponsor the monuments, which have ranged from small plaques at the base of trees to elaborate memorials. In the late 1990s the trend moved toward plaques on shared monuments.

During World War II the mighty Eighth Air Force could launch 3,000 combat aircraft on a single day. In October 1982 survivors and friends of the Eighth Air Force dedicated this monument to the unit's many achievements. The plaque beneath the propeller blade lists the names of seventeen members who earned the Medal of Honor. It also lists the number of other awards received, including 442,300 Air Medals and 7,033 Purple Hearts. The far side of the pylon has a map of southeast England, known as East Anglia, with the eighty air-fields of the Eighth Air Force.

Comradeship between air and ground crews during World War II is depicted in this memorial dedicated at the museum in September 1980. It was commissioned by former members of the 92nd Bombardment Group, a B-17 unit of the Eighth Air Force also known as "Fame's Favored Few." H. Richard Duhme, Jr., of Washington University handled the sculpting.

Here's where parents can rest (*above*) while children of all ages explore and touch things in the Discovery Hangar, located in the Modern Flight Hangar. The young visitor on the left, for instance, is learning the ABCs of Aviation by turning the giant pages. Special programs and classes are conducted periodically in the Discovery Hangar and elsewhere in the museum by the Education Division for children and their parents, and for teachers.

Over the decades a wide variety of civilian agencies have helped members of the armed forces and their families. Several illustrated panels describe those efforts in both war and peace.

Photographs, video testimony, and even artifacts smuggled out of Nazi concentration camps are part of Prejudice and Memory: A Holocaust Exhibit. The walk-through exhibit opened in early 1999. It reminds visitors of the six million Jews and millions of others who were killed by the Nazis. Some 185 downed U.S. airmen were imprisoned at the Buchenwald camp in WW II while attempting to avoid capture.

EDWARD C. GLEED FLYING JACKET

This World War II flying jacket belonged to Colonel Edward C. Gleed, USAF, a distinguished Tuskegee Airman. Colonel Gleed enlisted in the Army, assigned to Military Intelligence. In 1942 he entered and graduated in December as a ... the Euro-

Numerous flight jackets are displayed at the museum. This leather one was worn by Col. Edward C. Gleed, a Tuskegee Airman who officially shot down two enemy planes while flying P-51s with the segregated 332nd Fighter Group in Europe.

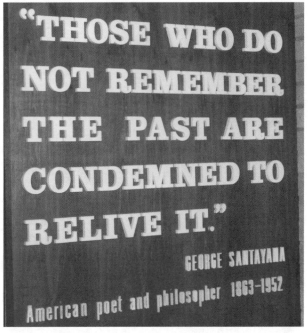

"THOSE WHO DO NOT REMEMBER THE PAST ARE CONDEMNED TO RELIVE IT."

GEORGE SANTAYANA

American poet and philosopher 1863-1952

These timeless words of wisdom from George Santayana underscore the works of this and many other educational museums.

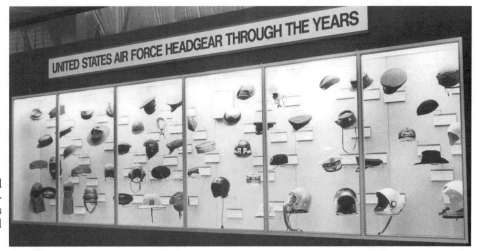

UNITED STATES AIR FORCE HEADGEAR THROUGH THE YEARS

Many former airmen from around the country have contributed headgear to this display. The pieces range from variations of old football helmets to modern space helmets.

Air Force people stationed at Wheelus Air Base paid for the stained-glass chapel window seen here. It commemorates the courage of the B-24 *Lady Be Good* crew who perished in the African desert in 1943. The window was brought to the United States when the U.S. Air Force vacated the Libyan base. Below are examples of Christian and Jewish chaplain traveling kits displayed at the sides of the window.

The explosion of this Russian-made rocket destroyed the U.S. Air Force chapel at Ton Son Nhut Air Base, Vietnam, on February 18, 196

Seal of the first Chief of USAF Chaplains, Chaplain, Maj. Gen., Charles I. Carpenter. (Donated by Chaplain, Maj. Gen., Charles I. Carpenter, USAF, Ret., Milford, Dela.)

USAF CHAPLAINS

The official origin of the chaplaincy within the American military services was a resolution passed by the Continental Congress in 1775. Since that time, military chaplains have administered to the spiritual needs of the serviceman, often at the risk of their own lives. For example, during WW II 8,896 ministers, priests, and rabbis served as chaplains with American military forces. Of these, 78 were killed in action, four died in Japanese prison camps, and 264 were wounded in action. The chaplaincy of the U.S. Air Force was separated from that of the U.S. Army on June 11, 1948 with the creation of the Office of the Chief of Air Force Chaplains, almost a year after the USAF became a separate service.

Today there are more than 1,000 chaplains within the Air Force, whose role involves not only worship and pastoral functions, but also religious and moral education, personal counseling, humanitarian services, cultural leadership, and public relations. Whether it be at a base chapel near a major American city or before an altar made of snow blocks north of the Arctic Circle, the chaplain brings solace to the lives of servicemen and women wherever they may serve.

The stained-glass window (*left page*) dominates the rear wall of the Chaplain Service kiosk, seen behind a B-18 wing. Displays on both sides of the walls relate the work of Air Force chaplains at home and abroad.

Protestant cross used in Chapel No. 1 at Randolph AFB, Texas from 1936 until 1972. (Donated by Protestant Chaplain[...] Randolph AFB, Texas)

This Protestant cross was used at Randolph Air Force Base, Texas, from 1936 to 1972.

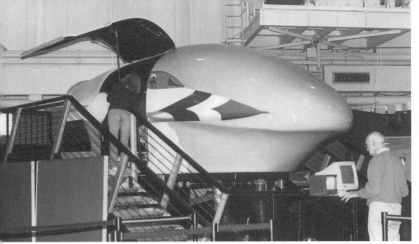

Adventuresome visitors in the Modern Flight Hangar can enjoy the sights, sounds, and motions of flight in this Morphis MovieRide Theater. The Air Force Museum Foundation charges a fee for the four-minute rides.

There is no fee for walking through certain display aircraft or to sit in the F-4 fighter cockpit (*right,*) or the F-16 cockpit simulator (*right, below*). The F-4 section is from the Phantom II that was used in the 1984 Call to Glory television series.

Mannequins are used throughout the museum to spotlight the work of Air Force members. Directly below, a young visitor gets a better understanding of an engine on a C-124 transport. Also shown are civil engineers under a C-46 transport building a WW II runway.

★ 9 | *Behind the Scenes*

There's more to the Air Force Museum than what is displayed for viewing by the general public. A great deal of activity goes on behind the scenes to acquire and prepare airplanes and other items for display. There's the guidance of the museum director and curator, of course. And then there is the detail work of four museum divisions: Collections, Research, Exhibits, and Restoration.

These divisions are located across the field from the main museum complex in several World War II structures. The first three are in Building 5, behind Hangars 1 and 9, while the latter is in Hangars 4C, 4D, and 4E. Because most of Wright Field is closed to the public, museum visitors must receive special passes to conduct business in these museum areas. However, escorted tours are conducted by volunteers to the Restoration Division on early Friday afternoons in June, July, and August. Registration is required.

The Collections Division stores portions of and maintains the inventory of the museum's vast collection in computers and on paper. This includes some 35,000 items in storage at Wright-Patterson AFB, 6,000 items on display at the museum, and 26,000 items on loan to museums, exhibits, and air parks at 360 locations around the world. Of the totals, about 300 aircraft and major missiles are at the museum and about 1,800 are on loan at other historical facilities.

Among the items in storage are aircraft, uniforms, parachutes, engines, armament, art, aircraft instruments, propellers, missiles, prisoner-of-war memorabilia, aerial photography equipment, communications gear, space items, flags, insignia, decorations, and trophies. All are available for use in new exhibits or, in some cases, for loan to other museums. In the case of the latter, the museum's Plans and Programs Division assists with on-site preservation guidance. Persons desiring to donate an item to the museum should first describe it in a letter to the curator.

The Research Division plays an important role when a "new" aircraft is being prepared for display or when a new exhibit is being designed. The Research staff can call upon its archives of more than eight million documents and 418,000 photographs for the detailed information needed to support the other divisions. This material also is available by appointment for qualified visitors with a scholarly or professional interest.

Museum staff members use data from the Research Division in a variety of ways. They determine what artifacts may be lacking but necessary to tell the Air Force story; then they attempt to locate those items. Information in the archives also provides guidance as to how an old aircraft should be refurbished or how mannequins should be uniformed. In addition, the Research people diligently seek out and write much of the explanatory material that appears throughout the museum's exhibit area. All of which aids greatly in the design and construction of habitats, or life-size dioramas, that starting in the late 1990s were drawing increased attention throughout the museum.

The Exhibits Division has the task of providing the detailed designs for a new exhibit and, supported by many hours of research, building the exhibit. This could be artifacts displayed in a wall panel, history portrayed in a free-standing exhibit, or a dynamic story told around a once-static aircraft.

One of the examples of the latter is the B-25 habitat that tells the story of the Doolitte Raiders of early WW II. Resting on a simulated carrier deck, the B-25 comes to life with nine mannequins portraying Lt. Col. "Jimmy" Doolittle, members of his crews, and crew members of the USS *Hornet* as they prepare to retaliate against Toyko for the Japanese attack on Pearl Harbor. Nearly 5,000 hours went into the project as each detail, from uniforms to inscriptions on ammunition boxes, was researched for accuracy. Visitors also can view a six-minute video from a 1980 interview with General Doolittle describing the mission. Several surviving members of the Doolittle Raiders and of the aircraft carrier crew, as well as Doolittle relatives, in 1997 attended the formal opening of the expanded B-25 exhibit.

The Restoration Division, as the name implies, restores aircraft and related equipment. Its four four-person crews also move aircraft as needed, suspend lighter ones from hangar ceilings for better viewing, manufacture replacement parts from wood or metal as needed for accuracy, paint aircraft to appropriately portray their role in history, rebuild badly damaged acquisitions, and generally strive on a daily basis to present some 300 aircraft and major missiles as genuine elements of aviation history. And when aircraft are forced of necessity to be displayed outdoors for a prolonged period, these crews must once again renovate previously restored aircraft. Staff members and volunteers also lend their expertise in replying to hundreds of inquiries on aircraft components from other museums and agencies.

When the museum in 1998 received President Kennedy's Air Force One, known as SAM 26000, Restoration crews dedicated countless hours into preparing the VC-137C (Boeing 707) for immediate public display. This included the installation of plexiglass walls along the aisles to preserve the interior cabin and its artifacts. Similar "plastic tunnels" were installed previously in President Roosevelt's *Sacred Cow*, President Truman's *Independence*, and President Eisenhower's *Columbine*. These four and five smaller presidential aircraft normally are displayed in the Presidential Hangar.

The behind-the-scenes work at the Air Force Museum, as in any museum, is never ending and entails the coordinated efforts of the entire staff and all volunteers. For example, special activities are conducted for school and youth groups by the Education Division, unique events for the general public are coordinated by the Public Affairs Division, day-to-day functioning of the museum facilities is managed by the Operations Division, and a variety of support is provided by the Air Force Museum Foundation Inc.

Many visitors first stop in the main lobby for guidance from volunteers at the Information counter. They know what's new and where to find just about anything. Information counters are also located in the Modern Flight Hangar, Eugene W. Kettering Gallery, and across the field in the entrance to Hangars 1 and 9.

Volunteers from the Officers' Wives Club (OWC) conduct tours for school groups during the academic year. Advance bookings are made by calling the Education Division at (937) 255-8048, extension 461 or 462. Here a guide tells students about the Doolittle Tokyo Raiders' use of the North American B-25B Mitchell bomber.

The Exhibits Division chief helps OWC volunteers prepare for Behind-the-Scenes Tours they conduct early Friday afternoon during summer months.

Rows and rows of files in this rotating Kardex contain Collections Division records of some 25,000 items on loan at 360 locations around the world.

No, they are not lost tourists or space travelers. These are mannequins waiting their turn for use in exhibits. Then they will be moved from the Collections Division to the Exhibits Division and into the museum.

During the accession process each new item is numbered (*left*), such as these boots from the Korean War. And there is a place for everything. Small insignia, for example, are stored in small boxes (*right*) in sliding storage trays that are kept locked. Thus when new exhibits are designed, authentic artifacts are readily available.

White gloves are required for much of the work in the Collections Division. Here the conservator prepares a condition report on a WW II bomber jacket before placing it into conservation.

When Air Force units are deactivated, their memorabilia are shipped to the Collections Division for storage until needed again. These heavy footlockers contain such mementos from aviation history.

This Spanish-built version of the German Messerschmitt Bf 109 was returned to storage in 1999 after the museum acquired an original Nazi WW II version. Since then it has been loaned to another museum.

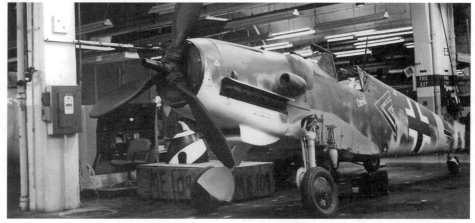

When documents are donated to and accepted by the Research Division, each item is processed in through a formal system. A volunteer is shown working with an archival collection donated by a former astronaut, retired Col. Mark Brown.

In order to prolong its life, a document such as this recruiting poster may be washed in purified water. This cleans it and removes any harmful acids. The next step could be encapsulating an historical document in mylar polyester for long-term preservation, either in the Research Division or in an exhibit.

The Research Division maintains millions of historical documents, technical manuals, aircraft drawings, photographs, personal papers, and other items related to Air Force history. As this volunteer would state, access to the materials is available to serious researchers by appointment only. Write to: USAF Museum, ATTN: Research Division, 1100 Spaatz St., Wright-Patterson AFB, OH 45433-7102.

Exhibits are planned in detail with computers, on paper, and sometimes augmented by miniature dioramas. Here in the Exhibits Division work is shown progressing on a tribute to Gen. Henry H. "Hap" Arnold, WW II chief of the Army Air Forces. The exhibit was built on a platform equipped with wheels for ease of movement to where needed in the museum for special occasions.

Miniature dioramas, such as this one for the museum's B-17G, help the Exhibits chief plan details for enhancing the flightline habitat scheduled for the Flying Fortress.

When replacement parts are needed for an artifact, they can be manufactured. Here a volunteer puts the finishing touches on a metal plate for a WW II bomb.

Early in the restoration process of this RB-47, volunteers inspect the tail section for damage suffered during years of outdoor display elsewhere in the country.

To reduce the weight pulling down on the wings of an aircraft and thus help preserve it, metal bombs can be replaced with those made from styrofoam and plastic. This Exhibits staffer is fashioning a substitute bomb from a block of styrofoam.

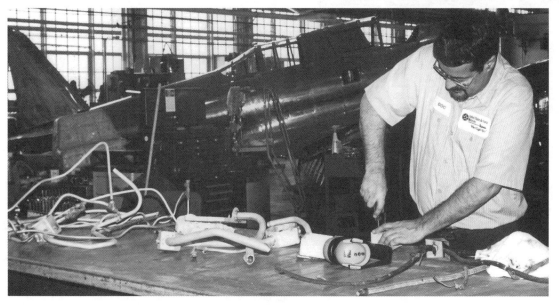

Parts for the instrument panel of the Northrop A-17A in the background are being worked on by this member of the Restoration team. The A-17 was the Air Corps' standard attack airplane in 1939, but was obsolete by the time of Pearl Harbor.

Supplies for a restoration project are discussed by a supervisor and a long-time volunteer who uses a wheelchair to help him with his work.

After being displayed out of necessity in the Outdoor Air Park, the Douglas B-23 Dragon and German Junkers JU-52 were towed indoors for restoration to preserve them for future generations.

Adjustments are made to a door leading into the cockpit of the Junkers trimotor. Adolf Hitler used a JU-52 as his private transport for a period in WW II.

Aircraft engines on display at the museum traditionally are restored by volunteers. This German Benz engine dates back to 1914-15 and the Albatross aircraft.

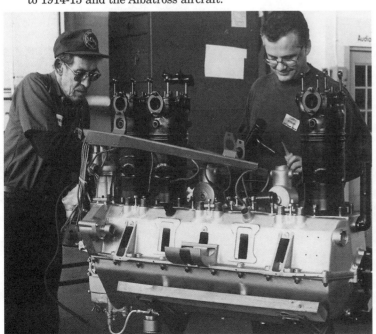

French designers produced the SPAD XIII C.1 under restoration to counter the German Halberstadt CL IV in WW I. Some of these SPADs were flown by the U.S. Air Service.

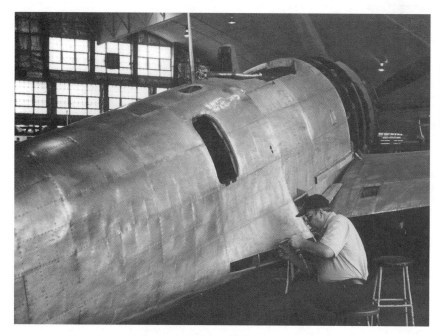

Even aircraft that have been displayed indoors for years are periodically restored, including this Kawanishi "George 21." The original builder's son, Tatsuya Kawanishi, president of Yamato Scale Co., in 1997 inspected the restoration project.

Having earlier restored several other Soviet aircraft, museum staffers at the turn of the century completed work on this MiG-23K, nicknamed "Flogger K" by NATO.

This portion of the Air Force Museum was dedicated in 1976 with actor James Stewart as the master of ceremonies and Senator Barry M. Goldwater as the keynote speaker. Both flew for the Army Air Forces in World War II. A large amount of the approximately $900,000 addition came from the estate of Brig. Gen. Erik H. Nelson, who died in 1970. He was a participant in 1924 in the first flight around the world.

As visitors approach the U.S. Air Force Museum from the main parking lot today, they view the 1991 addition (*left, rear*) and the Donors' Wall recognizing "Patrons of the Museum." These three monuments bear the names of individuals and organizations that have made major financial contributions over the years to the development of the museum. The Air Force Museum Foundation Inc. plans an expansion of the Donors' Wall as the museum itself continues to grow.

★ *Appendix*

All Department of Defense aircraft have been assigned designations to conform with joint Army-Navy-Air Force regulations.

Each aircraft or missile system designation has one letter to denote its primary function or capability: e.g., "B" for bomber, "F" for fighter, etc. To this, one or more prefixes are added to denote modified mission and status for aircraft, or mission and launch environment for missiles.

For example, in the designation VC-137, the basic mission or type is "C," cargo/transport. The "V" prefix denotes the modified mission of transporting staff personnel. If the designation were YVC-137, the additional "Y" prefix would denote prototype status. Suffixes are also used with aircraft designations to denote different models of the basic aircraft. Thus, the C-137B would be a newer version of the C-137A.

The following prefixes have been used for many years although not all apply to USAF aircraft.

In a missile system example, one model of the Minuteman ICBM is the LGM-30G. In this case, the vehicle type is "M" for guided missile. The prefix "G" denotes the mission, surface attack, and the additional prefix "L" gives the launch environment, silo-launched. If this LGM designation was prefixed with an "X," it would mean the missile system's status was experimental.

Status Prefix Symbols

AEROSPACE VEHICLES

Letter	Title	Description
G[1]	Permanently Grounded	Aircraft permanently grounded and utilized for ground instruction and training.
J	Special Test, Temporary	Aerospace vehicles on special test programs by authorized organizations, or on bailment contract, having a special test configuration, or whose installed property has been temporarily removed to accommodate the test.
N	Special Test, Permanent	Aerospace vehicles on special test programs by authorized activities, or on bailment contract, whose configuration is so drastically changed that return to its original configuration or conversion to standard operational configuration is beyond practicable or economic limits.
X	Experimental	Aerospace vehicles in a development, experimental stage where the basic mission symbol and design number have been designated, but not established as a standard vehicle for service use.
Y	Prototype	Aerospace vehicles procured in limited quantities, usually prior to production decision, to serve as models or patterns.
Z	Planning	Aerospace vehicles in the planning or predevelopment stage.

Modified Mission Symbols

AIRCRAFT

A	Attack	Aircraft modified to search out, attack, and destroy enemy land or sea targets, using conventional or special weapons. This symbol also describes aircraft used for interdiction and close air support missions.
C	Cargo/Transport	Aircraft modified for carrying cargo/passengers or medical patients.
D	Director	Aircraft modified for controlling drone aircraft or a missile.
E	Special Electronic Installation	Aircraft modified with electronic devices for employment in one or more of the following missions: (1) Electronic countermeasures. (2) Airborne early warning radar. (3) Airborne command and control, including communications relay. (4) Tactical data communications link for all nonautonomous modes of flight.

---

[1]Applies only to aircraft.

Letter	Title	Description
H	Search Rescue	Aircraft modified and equipped for performance of search and rescue missions.
K	Tanker	Aircraft modified and equipped to provide in-flight refueling of other aircraft.
L	Cold Weather	Aircraft modified for operation in the Arctic and Antarctic regions; includes skis, special insulation, and other ancillary equipment required for extreme cold weather operations.
M	Mine Countermeasures	Aircraft modified for aerial mine countermeasures and minesweeping missions.
O	Observation	Aircraft modified to observe (through visual or other means) and report tactical information concerning composition and disposition of enemy forces, troops, and supplies in an active combat area.
P	Patrol	Long-range, all-weather, multi-engine aircraft operating from land and/or water bases, modified for independent accomplishment of: antisubmarine warfare; maritime reconnaissance; and mining function.
Q	Drone	Aircraft modified to be controlled from a point outside the aircraft.
R	Reconnaissance	Aircraft modified and permanently equipped for photographic and/or electronic reconnaissance missions.
S	Antisubmarine	Aircraft modified so that it can function to search, identify, attack, and destroy enemy submarines.
T	Trainer	Aircraft modified and equipped for training purposes.
U	Utility	Aircraft, having small payload, modified to perform miscellaneous missions, such as carrying cargo or passengers, towing targets, etc.
V	Staff	Aircraft modified to provide accommodations, such as chairs, tables, lounge, berths, etc., for the transportation of staff personnel.
W	Weather	Aircraft modified and equipped for meteorological missions.

Launch Environment Symbols

ROCKETS AND GUIDED MISSILES

A	Air	Vehicles air launched.
B	Multiple	Vehicles capable of being launched from more than one environment.

Letter	Title	Description
C	Coffin	Vehicles stored horizontally or at less than a 45-degree angle in a protective enclosure (regardless of structural strength) and launched from the ground.
F	Individual	Vehicles carried and launched by one individual.
G	Runway	Vehicles launched from a runway.
H	Silo-Stored	Vehicles vertically stored below ground level and launched from the ground.
L	Silo-Launched	Vehicles vertically stored and launched from below ground level.
M	Mobile	Vehicles launched from a ground vehicle or movable platform.
P	Soft Pad	Vehicles partially or nonprotected in storage and launched from the ground.
R	Ship	Vehicles launched from a surface vessel— such as ship, barge, etc.
U	Underwater	Vehicles launched from a submarine or other underwater device.

Basic Mission and Type Symbols

AIRCRAFT

A	Attack	Aircraft designed to search out, attack, and destroy enemy land or sea targets, using conventional or special weapons. This symbol also applies to aircraft used for interdiction and close air support missions.
B	Bomber	Aircraft designed for bombing enemy targets.
C	Cargo/Transport	Aircraft designed for carrying cargo/passengers or medical patients.
E	Special Electronic Installation	Aircraft equipped with electronic devices for employment in one or more of the following missions: (1) Electronic countermeasures. (2) Airborne early warning radar. (3) Airborne command and control, including communications relay. (4) Tactical data communications link for all nonautonomous modes of flight.
F	Fighter	Aircraft designed to intercept and destroy other aircraft and/or missiles (includes multi-purpose aircraft also designed for ground support missions); for example, interdiction and close air support.
H[2]	Helicopter	Rotary-wing aircraft designed with the capability of flight in any plane; for example, horizontal, vertical, or diagonal.

[2]Type Symbols.

Letter	Title	Description
K	Tanker	Aircraft designed for in-flight refueling of other aircraft.
O	Observation	Aircraft designed to observe (through visual or other means) and report tactical information concerning composition and disposition of enemy forces, troops, and supplies in an active combat area.
P	Patrol	Long-range, all-weather, multi-engine aircraft operating from land and/or water bases, designed for independent accomplishment of: antisubmarine warfare; maritime reconnaissance; and mining function.
R	Reconnaissance	Aircraft designed to perform reconnaissance missions.
S	Antisubmarine	Aircraft designed to search out, detect, identify, attack, and destroy enemy submarines.
T	Trainer	Aircraft designed for training personnel in the operation of aircraft and/or related equipment, and having provisions for instructor personnel.
U	Utility	Aircraft designed for miscellaneous missions, such as carrying cargo and/or passengers, towing targets, etc. These aircraft will include those having a small payload.
V[2]	VTOL and STOL	Aircraft designed for vertical takeoff or landing with no takeoff or landing roll, or aircraft capable of takeoff and landing in a minimum prescribed distance.
X	Research	Aircraft designed for testing configurations of a radical nature. These aircraft are not normally intended for use as tactical aircraft.

Mission Symbols

ROCKETS AND GUIDED MISSILES

D	Decoy	Vehicles designed or modified to confuse, deceive, or divert enemy defenses by simulating an attack vehicle.

---

[2]Type Symbols.

Letter	Title	Description
E	Special Electronic	Vehicles designed or modified with electronic equipment for communications, countermeasures, electronic radiation sounding, or other electronic recording or relay missions.
G	Surface Attack	Vehicles designed to destroy enemy land or sea targets.
I	Intercept-Aerial	Vehicles designed to intercept aerial targets in defensive or offensive roles.
Q	Drone	Vehicles designed for target, reconnaissance, or surveillance purposes.
T	Training	Vehicles designed or permanently modified for training purposes.
U	Underwater Attack	Vehicles designed to destroy enemy submarines or other underwater targets, or to detonate underwater.
W	Weather	Vehicles designed to observe, record, or relay data pertaining to meteorological phenomena.

Type Symbols

ROCKETS AND GUIDED MISSILES

M	Guided Missile	Unmanned, self-propelled vehicles designed to move in a trajectory or flight path all or partially above the earth's surface, and whose trajectory or course, while the vehicle is in motion, is capable of being controlled remotely or by homing systems, or by inertial and/or programmed guidance from within. This term does not include space vehicles, space launch vehicles (space boosters), or naval torpedoes, but it does include target and reconnaissance drones.
N	Probe	Nonorbital instrumented vehicles (not involved in space missions) that are used to penetrate the aerospace environment and transmit or report back information.
R	Rocket	Self-propelled vehicles, without installed or remote control guidance mechanisms, whose trajectory or flight path cannot be altered after launch. Normally, rocket systems designed for line-of-sight ground fire against ground targets are not included.

AIRCRAFT INVENTORY

UNITED STATES AIR FORCE MUSEUM
Wright-Patterson Air Force Base, Ohio

Explanation of Inventory Codes

STATUS CODES

PV	On Public View; readily available in an exhibit building or the Outdoor Air Park.
R	Under Restoration.
S	Stored; generally not accessible for viewing.
W	Walk-through aircraft.
X	Displayed in markings not used on this particular aircraft when active.

LOCATION CODES

AP	Outdoor Air Park; adjacent to the main museum complex.
APG	Air Power Gallery; north section of front exhibit building.
EWKG	Eugene W. Kettering Gallery; third exhibit building.
EYG	Early Years Gallery; south section of front exhibit building.
MFH	Modern Flight Hangar; second exhibit building.
OB	Off Base; usually for restoration.
PH	Presidential Aircraft Hangar; Hangar 1 on other side of field.
R	Restoration; Hangars 4C-E on other side of field.
R&DH	Research and Development/Flight Test Hangar; Hangar 9 on other side of field.
RR	Outdoor Restoration Ramp; in front of Hangars 4C-E.
S	Storage; indoors, on other side of field.
*	Suspended or resting on partition at this location.

Codes above in some cases reflect temporary locations while construction was continuing on the Kettering Gallery. When completed, various aircraft will be relocated; this also occurs when restoration projects are completed.

The aircraft inventory was increased significantly in May 1998 when SAM 26000, the "Air Force One" made famous by President Kennedy, landed at Wright Field for transfer to the United States Air Force Museum. *U.S. Air Force photo by Gary Bays.*

Aircraft	Status	Location	Aircraft	Status	Location
A.V. Roe CF-100, Mk-iv	S	S	Convair F-102A Delta Dagger	S	S
Aero Design & Engineering	PV	PH	Convair F-106A Delta Dart	PV	MFH
L-26C Aero Commander			Convair C-131D	PV	AP
Aeronca L-3B Grasshopper	PV	APG*	Corben Super Ace	S	S
Albatross D-Va Replica	S	S	Culver PQ-14B Cadet	PV	APG
American Helicopter XH-26	R	R	Curtiss Model D Replica	PV	EYG
Beech AT-10	PVX	APG	Curtiss JN-4D	PV	EYG*
Beech AT-11 Kansan	PVX	APG	Curtiss O-52 Owl	PVX	MFG*
Beech UC-43 Traveler	PVX	APG*	Curtiss P-6E Hawk	PVX	EYG
Beech C-45H Expeditor	PV	PH	Curtiss P-36A Hawk	PVX	EYG
Beech T-34A Mentor	S	S	Curtiss Kittyhawk III (P-40E)+	PVX	APG
Beech QU-22B	PV	PH	Curtiss C-46D Commando	PVX	APG
Beechcraft VC-6A King Air	PV	PH	Curtiss-Wright AT-9 Fledgling	PV	APG
Bell P-39Q Airacobra	PVX	APG	DeHavilland DH-4B	PV	EYG
Bell P-59B Airacomet	PV	R&DH	DeHavilland DH-82A Tiger Moth	PVX	EYG*
Bell P-63E Kingcobra	PVX	R&DH	DeHavilland DH-89B Dominie	PVX	R&DH*
Bell UH-13J Sioux	PV	PH*	DeHavilland DH-98 Mosquito	PVX	APG
Bell UH-1P Iroquois	PV	MFH*	DeHavilland Canada U-6A Beaver	S	S
Bell X-1B	PV	R&DH	DeHavilland Canada C-7A Caribou	PV	AP
Bell X-5	PV	MFH*	Douglas A-1E Skyraider	PV	MFH
Bensen X-25A Gyrocopter	PV	R&DH*	Douglas A-1H Skyraider	S	S
Bleriot XI Partial Replica	PV	EYG*	Douglas A-20G Havoc	PVX	MFH
Boeing P-12E	PV	EYG	Douglas A-26A Counter-Invader	PV	R&DH
Boeing P-26A Replica Peashooter	PV	EYG	Douglas A-26C Invader	PVX	MFH
Boeing B-17G Flying Fortress	PV	APG	Douglas B-18A Bolo	PVX	APG
Boeing B-29 Superfortress	PV	APG	Douglas B-23 Dragon	S	S
Boeing B-29 Fuselage	PVW	MFH	Douglas RB-66B Destroyer	PV	MFH
Boeing WB-50D Superfortress	PV	AP	Douglas C-39	S	S
Boeing KC-97L Stratofreighter	PV	AP	Douglas C-47D Skytrain	PVX	APG
Boeing B-47E Stratojet	PVX	MFH	Douglas VC-54C Skymaster	PVW	PH
Boeing RB-47H Stratojet	PV	EWKG	Douglas VC-118A Liftmaster	PVW	PH
Boeing B-52D Stratofortress	PV	MFH	Douglas C-124C Globemaster II	PVXW	MFH
Boeing VC-137C	PV	PH	Douglas C-133A Cargomaster	PV	AP
Boeing EC-135E	PV	AP	Douglas O-38F	PV	EYG
Boeing NKC-135A Stratotanker	PV	AP	Douglas O-46A	S	S
Bristol Beaufighter Mk 1	S	R	Douglas SDB-4 (A-24)+ Dauntless	PVX	EYG
Caproni Ca-36	PVX	EYG	Douglas X-3 Stiletto	PV	R&DH
CASA 2.111H	S	S	Eberhart SE-5E	PV	EYG*
CASA 352L	PV	AP	ERCO Ercoupe 415	S	S
Cessna YA-37A	PV	MFH	Fairchild 24-C8F Argus(UC-61J)+	PV	MFH*
Cessna UC-78B Bobcat	PV	APG*	Fairchild C-82A Packet	PVX	AP
Cessna LC-126A	PV	MFH*	Fairchild C-119J Flying Boxcar	PV	AP
Cessna O-1G Birddog	PV	MFH*	Fairchild C-123K Provider	PV	AP
Cessna O-2A Super Skymaster	PV	MFH*	Fairchild PT-19A Cornell	PV	EYG*
Cessna OA-2A	S	S	Fairchild PT-26 Cornell	S	S
Cessna T-37B	PV	MFH*	Fairchild-Republic A-10A		
Cessna T-41A Mescalero	PV	MFH*	Thunderbolt II	PV	MFH
Cessna U-3A	S	S	Fairchild-Republic T-46A	S	S
Consolidated PT-1 Trusty	PV	EYG*	Fieseler Fi156C *Storch* (Stork)	PVX	APG*
Consolidated PBY-5A (0A-10A)+			Fisher P-75A Eagle	S	S
Catalina	PVX	APG	Focke-Achgelis Fa-330A-1		
Consolidated B-24D Liberator	PV	APG	*Bachstelze* (Sandpiper)	PV	APG*
Consolidated ZXF-81	S	S	Focke-Wulf FW-190D-9	PV	APG
Consolidated ZXF-81	S	S	Fokker D.VII	PV	EYG*
Convair B-36J	PV	EWKG	Fokker Dr.1 Replica	PV	EYG*
Convair B-58A Hustler	PV	MFH	General Dynamics F-16A		
Convair XF-92A	R	R	Fighting Falcon	S	RR

+Displayed as indicated within parentheses.

Aircraft	Status	Location	Aircraft	Status	Location
General Dynamics F-16A Fighting Falcon	PV	MFH	McDonnell Douglas F-15A Eagle	S	S
General Dynamics EF-111A Raven	S	RR	McDonnell Douglas F-15A Eagle	PV	MFH
General Dynamics F-111A Aardvark	S	RR	McDonnell Douglas-Northrop YF-23A	S	S
General Dynamics F-111F Aardvark	PV	MFH	Messerschmitt Bf 109G-10	PV	APG
Grumman J2F-6 (OA-12)+ Duck	PVX	MFH	Messerschmitt Me 163B	PV	APG
Grumman HU-16B Albatross	PV	R&DH	Messerschmitt Me 262A-1		
Grumman X-29A	S	S	*Schwalbe* (Swallow)	PV	APG
Halberstadt CL IV	PVX	EYG*	Mikoyan-Gurevich MiG-15bis		
Hawker-Siddeley XV-6A Kestrel	PV	R&DH	NATO "Fagot"	PV	MFH
Hawker Hurricane Mk IIA Replica	PV	EYG	Mikoyan-Gurevich MiG-17C		
Helio U-10D Super Courier	S	S	NATO "Fresco C"	PVX	MFH
Hispano Ha-1112-K (Mes 109G)+	S	S	Mikoyan-Gurevich MiG-19S		
Interstate L-6 Grasshopper	PV	APG*	NATO "Farmer C"	PVX	R&DH
Junkers Ju-88D-1	PV	APG	Mikoyan-Gurevich MiG-21F-13		
Kaman HH-43F Huskie	PV	MFH*	NATO "Fishbed E"	PVX	R&DH
Kawanishi N1K2-J			Mikoyan-Gurevich MiG-21PF		
Shiden Kai (George 21)	R	R	NATO "Fishbed D"	S	S
Laister-Kaufmann TG-4A	S	S	Mikoyan-Gurevich MiG-23K		
Ling-Temco-Vought XC-142A	S	RR	NATO "Flogger K"	PV	AP
Lockheed EC-121D Constellation	PV	AP	Mikoyan-Gurevich 29A	S	RR
Lockheed VC-121E Constellation	PVW	PH	Mitsubishi A6M2-21 "Zero"	S	S
Lockheed AC-130A Hercules	PV	RR	Nieuport N.28C-1 Partial Replica	PVX	EYG
Lockheed AC-130A Hercules	PV	AP	Noorduyn UC-64A Norseman	PVX	APG*
Lockheed VC-140B JetStar	PV	PH	North American A-36A Apache	PVX	APG
Lockheed C-60A Lodestar	PV	AP	North American B-25D		
Lockheed YF-12A	PV	R&DH	(B-25B)+ Mitchell	PVX	APG
Lockheed F-80C Shooting Star	PV	MFH	North American B-45C Tornado	S	R
Lockheed F-94A Starfire	PV	MFH	North American XB-70A Valkyrie	PV	R&DH
Lockheed F-94C Starfire	PVX	PH	North American F-82B		
Lockheed F-104A Starfighter	PVX	AP	Twin Mustang	PV	MFH
Lockheed F-104C Starfighter	PV	MFH	North American F-86A Sabre	PVX	MFH
Lockheed YF-117A Nighthawk	PV	MFH	North American F-86D Sabre	PVX	MFH
Lockheed P-38L Lightning	PVX	APG	North American RF-86F Sabre	R	R
Lockheed XP-80R Shooting Star	PV	R&DH	North American F-86F	RR	S
Lockheed SR-71A	PV	MFH	North American F-86H Sabre	PV	MFH
Lockheed NT-33A Shooting Star	PV	R&DH	North American F-100C		
Lockheed T-33A Shooting Star	PV	R&DH	Super Sabre	PV	APG
Lockheed U-2A	PV	MFH*	North American F-100D		
Lockheed-Boeing-			Super Sabre	PV	R&DH
General Dynamics YF-22	PV	MFH	North American F-100F	R	R
Loening OA-1A	PV	EYG	North American YF-107A	S	RR
Luscombe 8A Silvaire	S	S	North American O-47B	PVX	PH*
Macchi MC-200 *Saetta* (Lightning)	PV	APG	North American OV-10A Bronco	PV	MFH
Martin MB-2 Replica	PV	EYG	North American P-51D Mustang	PVX	APG
Martin B-10	PV	EYG	North American AT-6D Texan	S	S
Martin B-26G Marauder	PVX	APG	North American T-6C Texan	S	S
Martin EB-57B Canberra	PV	MFH	North American T-6D Texan	PV	MFH
Martin SV-5J (X-24A)+	R	R&DH	North American BT-9B Yale	PVX	EYG
Martin X-24B	PV	R&DH	North American T-28A Trojan	PVX	PH*
McDonnell F-4C Phantom II	PV	MFH	North American T-28B Trojan	PVX	R&DH
McDonnell RF-4C Phantom II	S	S	North American T-39A SabreLiner	PV	PH
McDonnell YF-4E Phantom II	S	S	North American X-15A-2	PV	R&DH
McDonnell F-4G Phantom II	S	RR	Northrop A-17A	PV	EYG
McDonnell XF-85 Goblin	PV	R&DH	Northrop YC-125B Raider	PV	AP
McDonnell F-101B Voodoo	S	RR	Northrop YF-5A Freedom Fighter	PV	MFH
McDonnell RF-101C Voodoo	PV	MFH	Northrop F-89J Scorpion	PVX	PH
McDonnell XH-20 Little Henry	PV	R&DH*	Northrop P-61C (P-61B)+		
			Black Widow	PVX	R&DH

+Displayed as indicated within parentheses.

Aircraft	Status	Location	Aircraft	Status	Location
Northrop T-38A Talon	S	S	Schweitzer TG-3A	S	S
Northrop T-38A Talon	S	S	Seversky P-35	PVX	EYG
Northrop AT-38B Talon	S	S	Sikorsky CH-3E	S	RR
Northrop Tacit Blue	PV	MFH	Sikorsky YH-5A Dragonfly	PV	MFH
Northrop X-4	PV	MFH*	Sikorsky UH-19B Chickasaw	PVX	MFH
Packard LePere LUSAC 11	PV	EYG	Sikorsky R-4B Hoverfly	PV	APG*
Piper J-3C-65-8 Cub	PV	MFH*	Sikorsky R-6A Hoverfly	PV	APG*
Piper L-4A Grasshopper	PVX	APG*	Sopwith Camel F-1 Replica	PV	EYG
Piper PA-48	PVX	R&DH	SPAD VII	PVX	EYG
Pratt-Read LNE-1(T6-32)+	S	S	SPAD XIII	R	R
Republic F-84E Thunderjet	PV	MFH	SPAD XVI	S	S
Republic YF-84F Thunderstreak	PV	R&DH	Sperry-Verville M-1 Messenger	PV	EYG*
Republic F-84F Thunderstreak	PV	MFH	Standard J-1	S	S
Republic XF-84H Thunderstreak	PV	R&DH	Standard J-1	PV	EYG*
Republic RF-84K Thunderflash	S	S	Stearman (Boeing) PT-13D Kaydet	PV	EYG*
Republic RF-84K	R	OB	Stinson L-5 Sentinel	PVX	APG*
Republic XF-91 Thunderceptor	PV	R&DH	Supermarine Spitfire Mk PRXI	PVX	APG
Republic F-105D Thunderchief	PV	MFH	Supermarine Spitfire Mk VC	PV	APG
Republic F-105G Thunderchief	PV	MFH	Taylorcraft L-2M Grasshopper	PV	APG*
Republic P-47D Thunderbolt	PVX	APG	Thomas-Morse S.4C Scout	PV	EYG*
Republic P-47D Thunderbolt	PVX	APG	Vertol (Piasecki) CH-21B		
Rockwell B-1A	S	RR	Work Horse	PV	PH
Rockwell B-1B	R	S	Vought A-7D Corsair II	PV	MFG
Ryan Navion 205 (L-17)+	PVX	PH	Vultee BT-13B Valiant	PV	APG
Ryan PT-22 Recruit	PV	APG	Vultee (Stinson) L-1A Vigilant	PVX	APG*
Ryan ST-A (YPT-16)+	PVX	PH	Waco CG-4A Haig	PV	APG*
Ryan X-13 Vertijet	PV	MFH	Wright 1909 Flyer Replica	PV	EYG
Schweitzer TG-3A	S	S	Wright 1912 Model B	PV	EYG
			Yokosuka MXY7-K1		
			Ohka (Cherry Blossom)	PV	APG

+Displayed as indicated within parentheses.

It took a bit of planning to move this Boeing B-52D into the Modern Flight Hangar. Museum staffers succeeded by turning its landing gear at an angle and towing the Stratofortress indoors sideways. They also dropped the hinged tail. Some other museum facilities are visible in the background. The double hangar (*left of center*) is known as the Presidential Aircraft Hangar and the Research and Development/Flight Test Hangar. The hangars to the far right include the home of the Restoration Division. *U.S. Air Force photo.*

Index ★ ★ ★ ★ ★ ★ ★ ★ ★ ★ ★ ★

+Displayed as indicated within parentheses.

More than 20,000 spectators witnessed the then last USAF landing of a Lockheed SR-71 when this "Blackbird" touched down in March 1990 on the runway at the south end of the main museum complex. The U.S. Air Force Museum also owns the SR-71s displayed at other museums around the country. *U. S. Air Force photo by Brian Barr.*